Memoir
Bath 2026

Gridlessness

Finding Freedom Off-Grid

Jeff Burkinshaw

Jeff Burkinshaw

Copyright © 2025 by Jeff Burkinshaw

All rights reserved.

No portion of this book may be reproduced in any form without written permission from the publisher or author, except as permitted by Canadian copyright law.

This publication is sold with the understanding that neither the author nor the publisher is engaged in rendering legal, investment, accounting or other professional services. While the publisher and author have used their best efforts in preparing this book, they make no representations or warranties with respect to the accuracy or completeness of the contents of this book and specifically disclaim any implied warranties of merchantability or fitness for a particular purpose. The advice and strategies contained herein may not be suitable for your situation. You should consult with a professional when appropriate. Neither the publisher nor the author shall be liable for any loss of profit or any other commercial damages, including but not limited to special, incidental, consequential, personal, or other damages.

Front Cover Design by Kerry La Porte

Back Cover Design by Kerry La Porte and Photography by Jeff Burkinshaw

Photographs and Artwork by Jeff Burkinshaw

First edition 2025

ISBN 978-1-0698566-0-9

Contents

Dedication	V
Introduction	1
1. Raw Land	5
2. Flirting with Rose	11
3. The Tarp House	15
4. Freedom 25?	29
5. Dirty Jobs	39
6. Brand New Baby and Ancient Technology	47
7. Headed North	55
8. Fire and Ice	69
9. Homesteads	79
10. Professional Hunter	95
11. A Bloody Joke	107
12. Bear, Elk and Bison	117
13. Dream Property	127
14. School is a Joke	135
15. Off the Grid	145
16. Cherryville Interlude	153
17. A Financial Crisis to Call Our Own	161

18.	The Hovel	171
19.	Toilets are a Big Freaking Deal	183
20.	Glorious Independence	189
21.	Off-Grid Systems	199
22.	Bonked Him Over and Chainsawed His Head Off	211
23.	The Great Flood	221
24.	True Wealth	231
25.	The Year Everything Happened	237
26.	Wandering in the Wilderness	251
27.	Freedom 35 and My Double Life	263
28.	The Open Ocean	275
29.	Living Like a Millionaire	287
Epilogue		297

For Rose. You are everything a wife and mother should be, an example of God's plan for humanity.

Introduction

We observed from a unique vantage point when the world went sideways in early 2020. Within the first couple weeks, we started getting curious phone calls from friends, family and more distant acquaintances. Mostly in jest, but serious enough to be a little awkward. They all had a similar message: "If things get any crazier, we're coming to your place!"

We had never been preppers but the truth was, between our root cellar, deep freezer and all of Rose's canning, we had at least a year's worth of food. Besides just sustenance, we had pretty much everything else we needed on our Off Grid Homestead. It was a sanctuary in the woods; peaceful and full of provision.

Normally, we went into town every week or so to pick up a few things, especially fresh fruits and vegetables in the winter. But we didn't need to. And with town getting super weird, we were easily able to go a few weeks without hitting a grocery store. If things got really crazy, we could dramatically change our normal eating habits and forsake town completely for a year or more. While that situation wouldn't necessarily be comfortable or convenient, knowing our lives were not at risk of global supply chain disruption, we had no need to panic.

Homesteading, being off-grid, and living at a distance from town always requires a certain amount of preparation.

When a hunter harvests a big animal, it could be months or even a year's worth of meat. When a farmer harvests a crop, it can easily last a year. Even us "hardly farmers" collected over five hundred pounds of potatoes from our earliest and most modest garden. With a good cellar, those potatoes can last until the next year's harvest. Having stored resources at home definitely took the edge off any apocalyptic concerns we might have had.

But more than that, we were all prepared in the most important way – mentally. We were used to taking care of ourselves. We were used to solving our own problems without a parts store or service tech just down the street.

We had skills.

We could raise food, hunt for food and forage wild foods. We could fix our own cars and equipment and build our own structures. And, of course, we could generate our own power, collect our own wood fuel, harvest our own rainwater and dispose of our own waste.

We bought toilet paper like everyone else, but if the end times actually came and cut off the supply forever, we had a near-infinite supply of..... pinecones!?

Maybe it was because of our physical and psychological separation from mainstream society but, from my angle, everyone just lost their minds. If you weren't afraid, none of it made sense. And if it didn't make sense, you were an outsider.

It was clear I wouldn't be working full-time in 2020, and that suited me just fine. We decided to invest our time improving our own self-sufficiency and enjoying what would be our last few months all together as a family.

Finding ourselves in this enviable position of self-sufficiency was not predictable. Rose and I had grown up like countless millions of others, sandwiched into neighbourhoods that were wedged into cities that merged into mega-cities. But we had a dream of experiencing life from the wide-open spaces.

We were not propelled off the beaten path by fear. We were not motivated by a need for isolation. We simply looked at the big picture, questioned the conventional, and then pursued the path of maximum freedom.

Of course, freedom comes with risks, mistakes and responsibility. Freedom offers great success, embarrassing failure and the distinct possibility of tragedy.

We have experienced it all. We've wallowed neck deep in muskeg, scrambled across crumbling ledges and conquered mountain tops. Our path has taken us from financial despair to windfall and from dirt-floor hovel to the quintessential cabin in the woods.

After leaving the safety of the suburbs, we chose a life of adventure. . . and life did not disappoint.

Raw Land

"Well, now what?" Rose asked, her eyes searching mine for a way forward. "Can we back all the way out of here?"

I was busy calculating our odds of death.

"You're not actually considering going through, are you?"

The odds of death were low, but not zero. I had been fooled before, with terrifying results. Still, the odds of success were enticing.

There was every indication we'd get stuck and stranded in the expansive body of water before us. Our map directed us to the gravel road, across the railway tracks and onto the "old logging road". We noted the remnants of a dilapidated homestead, made a few more turns and popped out of a forest of thick spruce into an abandoned pasture.

There was only one thing left on the map: the impassable puddle. Not just a puddle, or a big puddle, but an impassable puddle. Formidable. It ran into the timber on both sides and was maybe sixty or seventy feet long. And the depth? We had no idea. We definitely couldn't go around it and it appeared we couldn't go through it.

From the back seat of the four-wheel drive our five little girls adopted Rose's tone of concern.

A chorus of, "No! Don't do it!" and, "Daddy, I don't think we should!" erupted like a cacophony of baby birds.

Yet underneath the negative sentiments, I could hear the excitement in their voices.

"Can we take our seat belts off?" our oldest, Sarah, asked on behalf of all the sisters.

"Sure," I answered, knowing full well this would unleash all the pent-up energy of a nine, seven, five, three and two-year-old simultaneously.

Within seconds, all five girls were standing with their hands on the back of our seats, vibrating. The girls were a patchwork of sizes and colours, in an assortment of hand-me-down winter jackets, jeans and gumboots. The pigtails and ponytails bounced as all of us stared out the front windshield at the impassable puddle.

Rose started, "Jeff...," and I quickly hardened my heart to her mother's cry of reason as she continued, "We should just back up. It'll be fine. We can walk from there."

Well, no. We would not walk from there. With the screams and giggles I was hoping for, I revved the 7.3 litre turbo diesel and hurled my family and North America's largest SUV, the 8,900 pound Ford Excursion, into the impassable puddle.

It was early spring and the sun was trying hard, reflecting off the remnants of snow that glinted through the dripping conifer needles. The puddle was capped with rotten ice – blanketed with a pathetic layer of soggy snow and little puddles-on-top-of-the-puddle. Maybe, if we went fast enough, we could stay on top. The Excursion hit with a head of steam, scorching across the slush. A sense of optimism was at once conceived and then euthanized as the ice exploded and the truck dropped out from underneath us.

Rose gasped, clutched the dash, and let out a "Jeff!" that both blamed me and begged me. The girls unleashed a series of high pitched "Daddy!!" cries, each sister echoing the lead of her older sister. I responded with more throttle and felt the girls bounce off the back of my seat as the truck lurched into an underwater rut.

The puddle was clearly deeper than our front bumper, the ice making a painful racket as we crunched through it, like a clothes dryer full of hockey pucks. A tidal wave built in front of us as small icebergs flowed along each side of the truck. The wave continued to build as we neared the end of the puddle, first standing tall, then appearing to freeze in place, and finally crashing over the hood.

We spilt out onto dry land.

We were doing the impossible or, at least, passing the impassible. The truck came out like a wet Saint Bernard, wetter than the puddle itself. We all hopped out and the girls ran circles around the truck, inspecting for damage. The air smelled of hot dirt and exhaust and there was a distinct blast radius of mud, ice, and debris around the truck. The only real damage was our front license plate, which washed up on the edge of the puddle and was collected like a trophy by the girls.

"Jeff, are you happy now?" Rose asked with eyebrows raised.

I turned to Rose and we both grinned like giddy school kids.

"You KNOW I am!" I replied.

The first day on our new land was shaping up exactly as planned.

We had been looking for rural or remote land. We discovered forty acres totally surrounded by crown land, meaning there were no neighbours to contend with. It had no road into it and, when the map provided included an impassable puddle and then a mile long walk, we knew it was a gem.

It had to be close enough to town that I could still travel there for work every day, but far enough away that it was actually in the woods. We would have to open up the walking trail into a quad trail. Then, I could quad out to the truck, 4x4 to the main gravel road, and the rest would be easy. (A quad is like a four-wheeled dirt bike which northern Canadians use as a noun and a verb.) My commute would only be an hour. Eventually, we could build a new road and drive right on to the property. How special would that be?

We visited the property for the first time in early winter with our good friends, Dave and Shoshanna Godber, and their two boys. We had crossed paths just a couple years before and had instantly become close friends. We already owned land but were looking for something different and Dave and Shoshanna were eager to get onto their first property. We snowshoed into the property together and all agreed it was gorgeous – fresh snow, blue sky and moose tracks. It had me at "moose tracks."

Forty acres could easily be split. We would both be happy with twenty acres; especially as it was surrounded by miles and miles of crown land in every direction.

This property was otherwise unspectacular. Rolling land. Not too steep. And it had been logged about twenty-five or thirty years prior, so it had lots of young trees without significant timber value. It was mostly hard clay soil. It had no commanding vistas.

What it did have was isolation. And it could be ours. We agreed to buy it together, split 50/50. The Godbers would tame the wild west and we would command the east.

Jeff and Julia with Dave and Elliot scouting the new land

A month after our victorious surf into the property, the impassable puddle had become a normal part of our new routine. We cut open the trail to the property and had a couple old quads stashed in the bushes nearby.

Today was another big day. The girls bounced out of the truck, donned their overstuffed, oversized backpacks full of all their prized possessions, and Sarah and Abby, the oldest two, ran to the quads and fired them up. We had a trailer full of tools and building materials, but that would have to wait. First, we had to take possession of our new land. It was there, just a mile away down a muddy, slick, brushed-in trail.

Our wilderness empire awaited. Aside from the logging, no one had ever done anything to our land.

Not a thing.

Nothing was built on it. There was no trash. Nothing to clean up. Nothing to fix.

Just.

Raw.

Land.

It was what I had always dreamed of. We could build what we wanted, when we wanted, and how we wanted. And we would do it together as a family. Rose, on her beat-up, old Yamaha Kodiak with three little girls stacked high with backpacks, ball caps, jeans, and hiking boots, looked like she was made for this. And she was, but the truth is - even with a healthy serving of magic mushrooms - she could not have imagined herself in this position when we first met.

Flirting with Rose

Rose is gorgeous and stunning and kind and generous and patient, and somehow I coaxed her into a lifelong adventure with me.

She was quiet in school, but I noticed her the very first week of high school. In a class of one hundred and sixty early teens, she instantly stood out to me. At the time, like most thirteen-year-old stalkers, I mostly just drew pictures of her in my art book.

But the next year, in grade nine, I made a bold move. I asked her to the Christmas banquet. This was back when phone books were a thing and you could just look through it alphabetically to find someone's phone number. I didn't know her dad's name, so I had to phone a few different Pletts listed in the white pages, asking for Rosemarie, before I actually found the right number.

"Hello, is Rosemarie there?" I asked with an audibly high blood pressure and pulse rate.

"Yes, just a moment," responded her mom, followed by the muffled sounds of confusion echoing in a kitchen.

Eventually there was a "Hello?" with a slow and rising inflection.

"Hi, this is Jeff. From school... Uhm, I was wondering if you'd like to go to the Christmas banquet with me?" I managed, somehow even more awkwardly than any of my many practice runs.

"Hmmm." Pause. "Uh, maybe? Uhm, maybe I'll think about that and let you know," Rose answered.

Having executed the functional part of my communication plan, I realized I had no further plan and quickly jumped ship with a short, "Ok, bye!"

In hindsight, I guess it's a bit weird to ask someone on a date that you see almost every day but you've never actually talked to. Yet, at the time, it made perfect sense. If I was a bit more business savvy I would have known her response was a no-deal, but as an eternally hopeful fourteen-year-old, I patiently awaited her decision. The next week at school, walking to class, I was apprehended by her cousin. She cut in front of me, blushing deeply with sympathetic embarrassment and avoided eye contact.

"This is for you," she said and thrust a folded note at me.

I was confused until I noticed Rose and some friends watching from down the hall. She quickly looked away after catching my eye, and I knew at that moment the note bore bad news. With their anxious mission accomplished, the little circle of girls rushed out of sight, leaving me to open the note and read the verdict alone.

"Thanks for inviting me to the Christmas banquet, but I've decided to go with my friends."

Fair.

Objectively speaking, I was not her friend. I didn't know it at the time but, clearly, I was now playing the long game. I had just successfully put myself on the map.

It was a potato that helped turn the corner in our relationship, or non-relationship, depending on perspective. It was the summer between grade eleven and twelve when Rose took a job selling fruit and vegetables at a local fruit stand. It was a hot, ice cream, water-slide-holiday kind of day – and things were about to get a whole lot hotter.

Rose was wearing cut-off jean shorts and a navy-blue tee shirt. I was just hanging out in the shade of her fruit stand, admiring the view. With her ball cap and blonde ponytail, she whipped my teenage brain into a slurry, leaving me defenceless.

"Are you looking for something?" she asked. All business.

"I sure am," I thought, but replied, "How about one of these?"

I frantically scanned the bins, settling my eyes and pointing to a pile of potatoes.

"You want a potato?" she asked.

"Yes, just one."

She raised her eyebrows, smiled, let out a "Hmmm," grabbed the potato and rang it in. Not to let it become a distraction, I munched the potato while we chatted. It was the most delicious raw potato I've ever eaten.

Rose has that rare, stunning, natural beauty. It requires no fancy clothing, no makeup, no accessories. Long blonde hair with just a slight curl. Vivid, tanzanite blue eyes. Slender, but with perfect proportion. She's a ten out of ten, but what makes her a true knockout is her irresistibly radiant smile. She is the picture of beauty and joy.

As my good friend, Tom, says, "Potatoes never hurt no one."

True fact. I could have stood there eating potatoes all day.

The Tarp House

S ome things don't change.

Here I was, twelve years and five daughters later, staring at my beautiful wife. I was still stunned by her good looks, but her being here on this adventure with me took my adoration to a new level. This was not the kind of thing that had been on her bingo card. Part supermodel, part mama, and part commando in her work clothes and boots, she rolled down the trail packing half our kids and half our stuff. I took off behind her, thrilled to bits with our current circumstances.

The trail was now passable, but not without knowledge of the terrain and some serious effort. As it turned out, it was an unusually wet spring followed by an even wetter summer. The Ford Excursion had made it past the impassible puddle, but we now had another impassable puddle in front of us. The trail had about one hundred yards of low marsh and the first couple of times, it had been no problem. We floated on top of the vegetation and root balls without breaking through. But a few weeks and numerous heavy loads across that marsh transformed it into a smelly mud pit. No big deal for one of those fancy, high-powered quads but, chugging along with a bunch of extra weight, we just sank and churned.

We knew the drill. When the quad slows and the spinning tires shower us with mud, hop off and push it through. And we always made it – just barely. We arrived, time and again, covered in a special kind of aromatic, decomposing sludge. A literal path of adventure. Each laborious journey, and every obstacle we crossed, was like a brand-new toy under the Christmas tree.

There was no grand entrance to the property. No fence – really no indication at all – just GPS coordinates telling us we were now on the property. It was mostly indistinguishable from the forest.

There were pine, spruce, balsam fir and a few rare and precious Douglas fir. There were lots of poplar, some giant cottonwoods down at the creek and one or two birch trees, outliers of some birch groves in the forests nearby. There was plenty of willow if we wanted to weave a basket and enough alder for a lifetime supply of good smokewood. There were a couple of meadows, remnants of the old log sorting yards, that hinted at future pastures.

And there were caterpillars.

It was early spring and the leaves were just starting to unfurl. Amongst the vibrant green of the leaves, we noticed some beautifully-coloured, cute, little caterpillars. As we pulled up to the site of our future house we picked a few off our sweaters. They brushed off the bushes and crawled around innocently on our shoulders and our knees and fenders of the quad. We thought they were cute. We had no idea.

We spent evenings and weekends over the first four weeks cutting the trail and then getting to work on the homestead. We picked out a building site and started clearing land. The spot overlooked a small seasonal creek fed by a series of beaver dams off the property. It wasn't exactly a mountain stream but, to us, it was a fantastic water resource and a pretty setting. Within a few months, and after many failed attempts to cross the creek without capsizing our quads, we lovingly dubbed it Tipper Creek.

Our chosen building site was one of the flatter spots on the property and would give us the ability to build a house, shop and outbuildings on the same level. It was, however, an absolute tangle of young spruce punctuated with a few scattered poplars. The spruce were mostly ten to twenty feet tall but they were packed tight, each one within a few feet of five more. It seemed like we had to cut a hundred down just for our thirty-foot-by-thirty-foot building site. The stacks of fresh green spruce trees grew into small mountains as we cut and stacked, cut and stacked.

"Stand back, girls!" I said as I put the torch to the pile of felled trees.

Smoke and flames of biblical proportion shot skyward.

The forest fire danger was nonexistent due to the saturated ground, so we lit epic burn piles with impunity. We presided over raging columns of flames reaching fifty feet into the air, asserting dominance over our new landscape.

Everything we did seemed fantastic. Within days, what used to be a scraggle of immature forest turned into a wide-open, level building site. The massive pile of debris quickly turned into a warm glow of ash that, in its own way, helped clear a couple hundred square feet of land. We planned to build a little shop first and then, as soon as it was closed in, start building the house. Some solid shelter by Christmas was the goal, as northern Canadian winters don't mess around.

Of course, that didn't happen. We were about to get a lesson in contingency planning. The house we were planning would never be built.

The day we moved onto the land with Rose and the girls covered with mud and caterpillars, we set up a cheap pop tent, a folding table and designated a fire pit. It was a Friday afternoon and I had finished a full day of work in town, but the daylight hours were long so we still had a few hours before dark. The most pressing task was to hand-auger holes three to four feet deep to sink in our eight-inch-by-eight-inch cedar posts. These would form the foundation and structure of our shop. We had already laid out the grid, squared it and marked the post locations.

To start, we used an axe to chop the tangle of roots from the top six inches of dirt. Below that it was pure clay with just the odd large root to block progress.

We used a hand auger with a wooden T-handle that you rotate as you push down. It has a curvy metal basket at the bottom that cuts the soil and slowly fills up for extraction. Naturally, most of the soil just falls back down the hole. Not entirely satisfying, but reasonably effective. Maxed out, the auger would make a hole about seven inches in diameter, a little tight for our eight-inch-by-eight-inch posts. Once the hole was three or four feet deep we used a flat shovel to shave down the sides, enlarge the hole and make it square. To clean it out thoroughly I cut the pointy end off a little garden shovel, pressed myself into the ground, and strained to scrape out the remaining dirt a few ounces at a time.

While I was doing this, Rose was trying to set up beds in the tent for all seven of us, feed the girls snacks, take them to the potty (which, of course, we didn't have) and help the girls help me. The girls were always in the thick of it – never bystanders. The feeling of building my own home on my own land with my own family was, well, hard to describe. We were providing for our needs, seeing the result of our labour every day as we built something physically from the ground up. We witnessed our kids learning new skills as they soaked it all up. I suppose there were some dangers – hammers, axes, draw knives – but I figured as long as it wasn't me using the tool, they didn't have enough strength to seriously hurt themselves.

Rose was like a mother duck, running around, keeping her little ducklings from falling in holes or wandering off into the forest or falling into the fire. She was camp cook, homeschool teacher, construction site supervisor, special assistant to the lead carpenter and all the other roles of a wife and mother.

That night we got another hole dug and another post sunk. Things were really shaping up. As the long dusk set in, we admired our slowly rising shelter and more than one backwoods mosquito enjoyed its first meal of human blood. I resented the special attention I received from the wicked little creatures, so did not hesitate to shelter in the tent.

We had no water to wash up with. We had no electronics to play with. We didn't even have that much food to cook up. So, those first few days, life was pretty simple. We worked and then crashed in the tent. Then, we worked and crashed in the tent again.

And every day the mass of mosquitoes grew. We quickly learned that even in the heat of late spring we had to work with our entire bodies covered: long sleeves, gloves and hats with mosquito nets. Within a week, the mosquito horde dominated our fledgling construction project. They dictated all our activities. Eating became a new challenge. How do you eat with a mosquito net over your face? Going to the bathroom became a challenge. How can you stand to expose skin for any length of time? How do you bathe or wash? Well, that one was easy. You don't. We were learning to mitigate the onslaught of mosquitoes, but we didn't know they were just the welcoming party.

We had experienced mosquitoes before but had never battled black flies, and we were not prepared. If you've never had the personal pleasure, imagine a pint-sized, filthy house fly that prefers to creep around on your skin rather than fly directly to its desired location. Like a hateful ex exploiting your secret weakness, they seek out your most sensitive area and then chew at your flesh – the corner of your eyes, the insides of your ear canal, the crevice of your neck, and yes, if possible, they will seek out the most sensitive of your sensitive spots.

If you don't notice them crawling on you and brush them away, then you won't notice them when they start to chew. They have magical saliva that defers the pain until they're pretty much done hacking a hole in your skin and sucking your blood. It's often the case that you discover the wound only when you subconsciously scratch at the growing itch and find your fingers wet and sticky with blood.

Freaking black flies.

It was unfortunately common to find the kids working away at some portion of our new shop with random streams of blood running down their cheeks or out of their ears. I've got a lot of questions for the Lord and black flies are near the top of the list. Black flies? That's it. That's the whole question.

Because of their tendency to crawl, covering up doesn't really help, as they just crawl through the little crack in between your layers. They crawl up your shorts, if you are silly enough to wear shorts. They crawl up your shirt and under a mosquito net. Sometimes they buzz, frantically, inches from your face and then land somewhere. And you know they're in there, but unless you madly smack your own head and succeed in killing them, you just resign yourself to gnawing defeat.

Peak mosquitoes moved right into peak black flies and, although we fought valiantly, they left a lasting impression.

Not to be outdone by the flying, biting insects, a third plague soon overcame us. Turns out those small, colourful, cute caterpillars were mommy and daddy caterpillars. And they made approximately a hundred million tons of baby caterpillars. I'm not sure it made national news but it was definitely provincial news. The cyclical rise of tent caterpillars rose to historic proportions. They ate every leaf in the forest. The poplar trees, having just sprung vibrant, bright, little leaves, were quickly devoured.

Our regular quad ride on and off the property would leave us covered with caterpillars. When we arrived at the land or back at the truck the first thing we did was gingerly stand up and pick all the caterpillars off each other. If we didn't, we would inevitably squish them and crush them and smear them all over our pants and shirts and bodies and necks and the seats of our pants.

The caterpillars grew in number, until greasy slicks were observed on the highways where literally millions of caterpillars had crossed and been smashed into a nutritious spread. We had only been living on the property for two weeks: every day digging holes, planting posts and then beams, fighting insects, scraping off caterpillars, and then crashing in a big heap in our tiny tent. It was clearly a battle of attrition. Could the whittling down of our sheer body mass outlast the endless attacks of short-lived insects?

We didn't have time for second-guessing. The building progress was slow, we had limited time and resources, and a more substantial problem was brewing.

Our optimistic schedule depended significantly on the construction of a new road into our property. The owner we had bought the land from had a small logging outfit and included a fresh, new driveway as part of the purchase. The proposed route would push straight through from the gravel road, bypassing the mile-long quad ride through the muskeg, circumnavigating the impassable puddle and the abandoned homestead trail.

Once the road was in, we'd be able to drive our truck to the land and bring in large amounts of material and maybe some small equipment. But clearing the new right-of-way was painfully slow. It was a one-man operation, and that one man wasn't there very much. When he was there, he seemed to have the wrong equipment. After the proposed route was logged, we found him trying to build the road with just a skidder. Famously, skidders are for skidding logs out of the bush, not for building roads. It was clear he wasn't going to pull any stumps, dig any ditches or shape a road with a skidder.

But the biggest problem was that it rained and rained and rained all summer long. The equipment couldn't move in the oozing mud and, when it did, it mostly just made big, muddy holes. Eventually, he informed us that we'd received so much rain that no amount of future warm weather would dry it up before winter. If we were lucky, our first road would be an ice road.

Back at the homestead we took stock of the situation. I had taken two weeks off work with the hope of getting the shop closed in. At this point, half our posts and beams and rafters were put together. We had the skeleton of half a shelter – a soaking wet half of a shelter. There was not going to be a road anytime soon and so progress was not going to accelerate, especially with me going back to work.

It was at this point that we realized the shop we were building would need to become our house, for we certainly would not have time to build a shop and a house before winter at this rate. We hatched a new plan. I would take my last week of holidays, quickly close in our half of a house, and move in. It was early June, peak mosquito season, the ground had reached its saturation point, and we had seven days to get it done.

On top of our cedar posts we placed six-inch-by-eight-inch spruce timbers: three parallel rows of posts and beams, the middle being just a couple feet higher than the outer two.

First week on the land, Abigail with a moose antler in front of our pop tent

Christina, with the draw-knife, and Keziah helping Jeff peel logs

The rafters were made of raw logs from our property. There was a stand of mature dead pine trees, killed by pine beetles, about four hundred yards from the house. As a family, we headed up there on the quads and

I cut down a tree, delimbed it, cut it to about eighteen feet long and chained it to the back of the quad to be dragged back to the house. There was no smooth trail. In fact, there was no trail at all until we made a few passes.

Driving the quad and dragging those logs was our own version of the "Little Britches Rodeo." Every time the log jammed a stump or a tree trunk, the quad would rear up and then buck as it let go. Nine-year-old Sarah drove the quad and I ran along like the rodeo clown to guide the log over obstacles.

"Stop!" I yelled, as the butt end bore down on a stump.

"What!?" she turned towards me and yelled back over the rumble of the four-stroke.

Too late. She was tossed unceremoniously into the air as the log jammed the stump, jerked the chain and reared up the quad.

Back on the seat, she gripped the handles tighter and yelled again, "What!?"

I had already rolled the log clear and was forced to answer, "Nothing. Keep going!"

This entire, poorly-coordinated effort would inevitably play out over and over on each trip.

The rafters were way too heavy for us to lift up to roof height, so we built a crane. The crane consisted of a sixteen-foot log raised up alongside one of the posts with a dolly wheel lag-bolted to the top. We tied a rope to the rafter and passed it up over the wheel, which acted like a pulley, and then down to the quad winch. Rose and I needed to handle the log and try to fit it on the beams, which meant Sarah had to operate the winch. The combined weight of Sarah and the quad was less than the rafter logs, so the quad had to be chained to a tree behind them. Even at that, the front of the quad would lift right off the ground.

The very first rafter was the trickiest. Aside from a few braces, there was not a lot of stability. Our log-crane butt-end just sat on the dirt, supported midway up, chained to the same post I was standing on. It was a little hairy but, thankfully, just ten feet off the ground.

"Ok, Sarah. You ready?" I asked, my eyes scanning our little construction site.

"I think so?" she offered without much conviction.

All the rigging seemed in place, and I double-checked there was a jacket tied around the winch line.

"Winch it in, Sarah. Take up the slack."

The ropes, chains, posts and beams all popped and twanged and creaked at the tension. I watched Sarah intently, and as the front of the quad shifted and began to lift I shouted, "Ok, Stop!"

"Jeff, are you sure it's ok?" Rose called from her perch on the far beam.

"Yep, it's fine," I responded, and then looked everything over to see if it was all fine.

There were two primary risks. The entire post and beam structure could fall down, including the one I was standing on, or the cable could snap and hit someone – probably Sarah. I was obviously indestructible, even if I went down with a jumbled pile of beams, but Sarah needed extra protection. The jacket on the winch cable should slow it down if it let go, but it could still do damage.

"Sarah, duck down so you can't see the cable, ok? Just listen to my instructions, ok?"

"Ok Dad!" she said, scooching down on one side of the quad.

I glanced at Rose and then gave the word, "Winch it in, Sarah!"

The log shuddered and rose slowly while the "crane" complained under the weight.

"Keep it going, keep it going. . . aaannnnnd stop!" I called out, keeping my eye on the massive weight dangling at my feet.

"Rose, come over here and pry down on that end and I'll pull this end up on the beam. Ok, Sarah, winch it just a bit. . . Ok, stop!"

Our scheme worked to lift the rafter off the ground and get it to a certain height, but it couldn't move it side to side. We had to muscle it the rest of the way. With Rose on one beam and me on the other, we played pass-the-huge-log-around and tried to land it on the wobbly beams we were standing on. It was like playing giant Jenga on a big bowl of Jello. After we landed that first rafter, it became considerably easier as the whole structure tied together.

On a good day, we could get two rafters harvested, dragged back to the house, flattened on the top side and lifted in place. While we were developing some efficiency, we were also running out of precious time. We had to close it in now.

Half the structure was complete, a thirty-foot-by-sixteen-foot skeleton, and in those desperate times, it called for three, heavy-duty hay tarps from Princess Auto. Yes, we were going to live in a tarp house. At this point, Rose didn't care. The thought of 480 square feet of inside space was pretty inspiring.

On the two side walls, we installed one-inch plywood along the bottom four feet and window screen across the top four feet. The gable-end walls and roof each received one giant, grey tarp. We pulled it over and stapled it down with a hammer tacker. I felt terrible putting all those staple holes through a perfectly good tarp, knowing that they would eventually leak into our brand-new abode. But this was a time for action, not sympathy or finesse.

Rose and the girls with half the house tarped in - the Tarp House

The gable-end walls were the most ridiculous, spanning sixteen feet between posts, the tarps dragging with shame on the muddy ground. The whole house looked like a redneck Christmas present wrapped in duct tape. And so, after living for three weeks in our sixty-four-square-foot pop tent, we moved into the 480-square-foot tarp house. One of the biggest upgrades of our life.

Let me take a moment to clarify why we went off the grid. Our main goal was a lifestyle free from debt, closer to nature with the freedom to enjoy a healthy, productive homestead. But was the off-grid part necessary?

Objectively, no. Yet the values and benefits of the off-grid part of our adventure would be difficult to separate from the rest.

First, off-grid property is generally way cheaper than a property serviced by the electrical grid. For this reason alone, I'm not sure we could have achieved the same results with the same amount of money.

Second, we wanted a truly wild property – in this case, completely surrounded by crown land and ten kilometres to the nearest neighbour. There was no grid in places like this. We couldn't live in a truly wild location and have grid power.

Third, to be off-grid was to be independent and to develop our own skills and systems. Partly, it's just wise to become more self-sufficient, but it's also just a passion of mine to understand and build things. In some cases, we did things the hard way – not necessarily because it was the best use of our time and money, but rather, because I just wanted to try something different.

And there is another, maybe less tangible reason, but still something we did intentionally. That is, we deliberately reduced our level of comfort and convenience. In the grand scheme of the world through time and in different regions, it is hard to overstate how comfortable and convenient our lives are – and, unfortunately, not always for the better. Like gluttony in the presence of too much food, it is hard for us to not take our wealth for granted and become ignorant and ungrateful. We become ignorant of how much effort, innovation, hard work, and energy are actually required to be able to keep ketchup cold in a fridge, and are ungrateful for everything we have, simply desiring more.

I wanted my family and myself to develop real insight and appreciation for our tremendous position in history and geography. It is a position where a middle-class family in Canada has more wealth and luxury than 99% of all people who have ever lived.

Freedom 25?

I hatched my Freedom 35 plan (originally a Freedom 25 plan) when I was seventeen years old. It was the summer between grade eleven and grade twelve. The summer of the potato. My buddy, Joe Falk, got me a job with his family's construction company. They did all sorts of agricultural, commercial and residential construction but, in my experience, we mostly worked on chicken barns. More specifically, I spent a lot of time renovating fifty-year-old chicken barns, which is akin to polishing stadium-sized turds. I didn't know it, but I was on a specialized career path in farm effluent.

One of the first projects I worked on was an effluent flushing system where we framed up a large concrete sump tank that would eventually house the discharge from the barn. With barely a tool to my name, it was with humiliation that I asked Ken Falk (my friend's dad and company owner) to borrow his hammer.

We were nailing together the final touches of formwork – two-by-fours and plywood – and tying rebar along the top edge. I was feeling pretty good when the concrete truck showed up. Imagine becoming an expert in framing, rebar-tying, and pouring concrete all in my first day. Just a few more nails and I would have the top support locked in. That's when, swinging with way more enthusiasm than skill, the boss's hammer caught the rebar with a twang and sprang from my hand, cascading musically to the bottom of the eight-foot form. The sound caught my supervisor's attention. We looked at each other and then we looked down at the hammer.

He raised his eyebrows and asked, "Is that Ken's?" and then laughed enthusiastically before I had a chance to answer.

The hammer lay a long way down through a maze of rebar and tie wire, and the concrete truck had just shown up and was already starting to pour the wall. I don't remember doing anything except staring first at the hammer, then at the supervisor, then at the chute of the concrete truck as it unapologetically buried the hammer forever.

Luckily, Ken is a fantastic guy. When he found out, he furrowed his eyebrows as long as he could, waiting for my shame to surface, then burst into laughter. A good boss, like a good journeyman, is slow to anger yet takes great pleasure in the early failings of a trainee or apprentice. Smugly observing the emotional distress of a young worker, and then graciously offering mercy, is well worth the price of a hammer.

I learned a lot that day, and almost every day brought me new skills and abilities. I will always remember a helpful scolding I got from Grandpa Falk one day as he observed me apparently underperforming. I was struggling to maneuver a wheelbarrow full of concrete up a ramp when he came and took over.

"You just have to get..." he said, grabbing the wheelbarrow and gearing down, "more VIGOROUS!" he finished, as he thrust the cart up the ramp.

But the big epiphany for me that summer was that I could build stuff. It's totally amazing how a little skill can quickly turn a lift of two-by-fours into a long-lasting, functional structure. And how simple materials – rocks, cement and water – can build a foundation that will last for centuries. I would eventually go on to study electronics and electrical engineering, but the concept of building life-giving shelter with relatively simple methods and simple materials is still one of the most profound lessons of my life.

As we continued the project, we framed walls and a roof, clad it all with roofing metal and closed in the sediment tank. The whole structure went up and was sealed in within a week. I was so inspired. I could live in there. It was substantially bigger than my bedroom, and how much bigger would it need to be? There's no reason I couldn't build my own house with concrete, two-by-fours and sheet metal, and it would just take a few thousand dollars.

At this point in my life, I had ideas but I did not yet have a plan. My ideas involved marrying a beautiful woman, moving into the woods, hunting and fishing for most of my food and inventing stuff just for fun. None of my ideas included a career or a mortgage. In fact, I was genetically opposed to debt.

My Grandma and Grandad, born in the 1920s, had lived through good times and tough times and, with the weight of their experience, vehemently opposed debt. Their first house was assembled from an abandoned chicken coop and put together at the mill yard where Grandad worked. It was exceptionally modest, but it was free housing and it didn't seem to hurt them any. They would go on to have a life of adventure and always owned their own home, debt-free.

I don't remember an origin story for their debt-free convictions, but I believe it came primarily from biblical teachings such as Proverbs 22:7, where it talks about the borrower being a servant to the lender. In any case, I embraced this philosophy and regarded the lack of debt to be a prerequisite for a good life.

I always thought it was crazy that to buy a house, even a little, old house, it costs hundreds of thousands of dollars. And yet, looking around the Greater Vancouver area of British Columbia where I grew up, for the life of me I couldn't see anyone living in a repurposed chicken coop. That's why the chicken poop sump shack was such an epiphany for me. I now had hard proof that a house did not have to cost hundreds of thousands of dollars and, from that point, my life plan started to come together. I would buy land and build my own house.

My four-year campaign to win Rose's heart was finally successful. In the last days of the potato summer, at the first party for our graduating class, Rose and I crossed paths. I always got excited at big gatherings and bounced around connecting with as many people as I could. I made my way over to Rose's cousin (the same one who handed me the rejection note three years earlier) to say "Hi." We would be on the student council together for our grade twelve year. But as I approached, I noticed Rose standing alongside her. I turned to Rose, gave her a big smile, said "Hi" and then froze. Time stopped.

She melted.

She looked right at me with that million-dollar smile, her eyes sparkling and dancing like the Indian Ocean. She looked right through my eyeballs, right through my brain and right into the core of my body. While her emotional response was breaking the needle off seismic detectors, her verbal reply was simply, "Hi. . ."

With our eyes locked, and without taking a breath or speaking a word, I silently told her that I had always loved her. She seemed to understand. "How are you?" I offered, quietly.

"I'm good," she said.

She was still staring right at me and her eyes seemed desperate to share all her hopes and dreams with me. Her cousin said something but neither Rose nor I responded and she walked away in disbelief.

"Do you want to hang out?" I asked.

Rose nodded, and we spent the entire evening together. It was instant. A goal in sudden-death overtime.

Rose was mine.

And we have been together ever since that day.

Grade twelve started with half a plan and a beautiful girlfriend.

I think, like most high school sweethearts, we often imagined being married. I remember telling Rose once that all I wanted to do was hunt in the woods like a savage mountain man, kill a wild beast, drag it home and put the carcass on her doorstep. Rose thought I was just being cute at the time but eventually would look back and realize how close my vision was to our reality.

As high school ended and the real world started, Rose and I quickly began to calibrate our future together. We had to take our vision of life together and turn it into the nuts and bolts of a working machine. We were going to need our freedom plan sooner rather than later.

It was a bit of a dilemma. You see, I had a gorgeous, faithful, kind, supportive girlfriend, and I didn't know exactly what to do with her. I was seventeen, and yet, it was time to make the biggest decision of my life. Having a girlfriend was great, but I wanted to get on with a life of adventure and freedom and productivity. I wanted to do stuff and build stuff, and I would either start down that path by myself or I would need a wife. I had never really anticipated having a wife so quickly, but I also had never imagined a life on my own.

I had to use my brain.

Was she a truly good woman? Were her family quality folk? Would she be a joyful and supportive partner? Would I regret not securing the best wifely candidate I'd ever seen?

The answer to each question was "Yes" and so, four months after graduating high school and two months after turning eighteen, I bought a diamond ring. We were engaged in the spring, planned a summer wedding, and drafted a strategy.

I would go to tech school and get some skills so that I could get a cool job inventing stuff. Rose was finishing up a one-year legal office assistant course and would keep us afloat for the next couple years.

I had started the first semester of a science degree program but was immediately dissatisfied. It was just like high school: a lot of theory, a little bit of carefully controlled and strictly unoriginal lab work, and, most disappointing, no building or inventing.

Instead, for the eight months leading up to our wedding, I continued my illustrious career of chicken barn effluent system rehabilitation. During that time, we also built exactly one house from excavation to lockup. As we sealed up that house in the spring of 1999, Rose and I agreed on our long-term goal, Freedom 25. In seven years we would have skills, a productive piece of land, a house we built ourselves, and be a hundred percent debt-free. We would be free to pursue a life of adventure and purpose without the ball and chain of a mortgage.

One glaring oversight in our plan that you might have noticed was children. As an eighteen-year-old, I had never credibly imagined being a father. It's not that I didn't want to, it's just that I had never thought of it. This lack of foresight would become apparent very quickly.

At this point it might be good to make a distinction of sorts. The terminology for our Freedom 25 plan was borrowed from a popular series of TV commercials advertising investment products during the 1990s. For some reason, although several decades away from their target audience, I was impacted by their Freedom 55 pitch. The promise of the Freedom 55 retirement plan was financial independence. If you worked hard enough and long enough and invested enough and got a good enough return, you would have one million bucks in the bank when you hit fifty-five years of age. You could live off the interest and golf, vacation and generally please yourself for your "golden years."

Teenagers

At that time, I naively believed freedom was directly related to financial independence and, if you still needed to work, you were not free. Although I dispute this assertion strongly and with a multifaceted argument now, my youthful perspective was slightly misled by the retirement philosophy. Thankfully, it did not derail our early Freedom Plan as we developed our eventual winning strategy.

I never fought against the idea of hard work. Although I could never work as hard as Grandad, who genuinely enjoyed manual labour, I believed in it, and have always believed it should be embraced.

Grandad got his first real taste of freedom when he was fourteen.

It was 1940 and he was on a ship in the middle of the Atlantic Ocean without parents or any other guardians. For the first time in his life there was no one to tell him what to do. He had no teachers or boss to direct his comings and goings. On the Duchess of Bedford with him were 1,250 other children from England, all fleeing the feared decimation of the motherland. He roamed that ship for ten days as it rocked and rolled across the ocean, and he loved it. What an adventure.

When the Duchess eventually docked in the Saint Lawrence at Montreal, he boarded a train and continued for another five days, unsupervised. It was a monumental journey westward, from Oxshot on the outskirts of London, across the Atlantic, across Canada to Prince Rupert and then by small boat up to Georgetown Mills. Never had he imagined seeing so much of the world and so much of the wild, Canadian frontier.

He settled on the rugged and remote north coast of British Columbia, neighbouring Alaska. Having conquered the Atlantic, he now set his sights on the Pacific. For the next five years, he worked for his Uncle Stub at the box mill and saved up his money. Still a boy, he explored the coastline endlessly and hiked the creek up to Georgetown Lake to fish for cutthroat. He spent as much time on the ocean as he could, fishing salmon and halibut.

At sixteen years old, he sold his victory bond and invested $500 in his own freedom plan – a twenty-six-foot wooden trolling boat, the Alice

M. It wasn't a house, but it was a place to live, and it would allow him to eventually take up salmon fishing full time. He explored and learned to ride out the wild northwestern storms. He was debt-free, had a place to live, and was doing work he loved.

Only days after quitting the mill, moving all his possessions onto his boat, and starting his new sea-centric life, it all came to an explosive end. After spending the night on shore, he was jarred awake by a friend in the early morning to see a perfectly innocent wisp of smoke coming from the cabin of his boat.

It was a perfect tragedy.

He scrambled into his friend's boat and approached the Alice M anchored in the bay, but by then the wisp of smoke had turned into licking flames. He remembered the twenty-five gallon gasoline fuel tank. Could he get to the boat and put out the fire before it reached the tank, or would he climb onboard just as it exploded? With wisdom and sadness, he returned to shore just as the fuel tank ignited and blew apart the Alice M in a violent fireball. He stared for hours as the hull burnt down to the waterline, drifted to shore and smoldered in the mud.

His fishing gear, hunting rifle, clothes, wallet – everything he owned – was properly destroyed by fire and water. His audacious plan to live modestly without debt and pursue work that he enjoyed and found meaningful had failed. With his head hung low, he returned to the box mill and humbly asked for his old job back. He needed to rebuild his savings and revise his plans.

Dirty Jobs

I may not be Grandad, but I've done some heavy lifting and I've worked a few long days in less-than-ideal conditions.

One Christmas break during high school, I got a job helping to install a new metal roof on an old chicken barn. It was ideal roofing weather – consistent rain and just above freezing. I owned the cheapest rubber rain suit available – the kind that collects every ounce of your perspiration, condenses it on the cold, rubber inner lining, and generously donates it back to your clammy skin.

I was a seventeen-year-old labourer. Nothing in my job description promised comfort. So, despite feeling hot and cold and squirmy all at the same time, I was happily slogging along. We had been moving materials and preparing the roof, but it wasn't until we started lifting sheets onto the roof that things became memorable.

"How about I climb on the roof and you stay on the ground to pass the sheets up to me?" my coworker offered.

"Sure," I replied, a little too quickly.

The sheets of corrugated, low-profile roofing metal were about twenty feet long and three feet wide. I would grab each piece with a hand on each side, letting it sag in the middle like a twenty-foot-long trough. The parabolic shape made the otherwise flimsy metal into a temporarily rigid structure, allowing it to be rotated from horizontal to vertical and passed up to the worker on the roof. He would then pull it up to the peak while I supported it from below. I didn't know it then, but it was this exact task that encouraged me to pursue post-secondary education and a skilled career. In fact, there were many times during particularly exhausting periods of study that I would look back to this exact moment to rediscover my motivation.

The rain poured.

As I lifted that first piece and supported the bottom edge of a sixty-square-foot water-collecting masterpiece, I realized I had been outsmarted.

"Hurry up, if you can!" I begged.

I had been sweaty, clammy and content in my rubber rain suit but now, with arms outstretched above my head and five-gallons-per-minute of freezing water entering each sleeve of my rain jacket, I was legitimately miserable. My sly coworker took his time driving screws through the metal, securing the panel with deliberate tenderness. Like a deep-fried ice cream sandwich, the combination of rubber-lined sauna and Wim Hof ice bath was sensational. If it had just been a passing condition of a particular task, it would have never made an impression, but repeating that act of self-deprecation all day long cemented my resolve.

I've also done some truly disgusting jobs. In fact, I may have performed the trifecta of dirty jobs. If you're a fan of Chinese food, you might want to skip this one.

I was on a commercial construction site during the earth and civil works of a new development behind an old strip mall. It was a tight space between the new cinder block wall and the old building, too narrow for any type of skid steer or excavator. I would be the machine for this job.

The task was simple. I just had to level and grade the rocky soil between the two concrete foundations. Except it wasn't all rocky soil. It turns out a Chinese restaurant had, for decades, emptied its old grease traps out the back door. Previously concealed by the overgrown vegetation of a vacant lot, the new excavations exposed toxic sludge oozing like pus from a bursting abscess.

My steel-toed boots were wholly inadequate as I waded in to make sense of the man-made tar pit. A shovel at a time, I dug through layers of fresh soil, native soil and sludge. I eventually found the bottom. It was exactly six inches over the top of my boots. I spent ten hours with a shovel, digging down protruding islands of disturbed soil and then smearing them with the sour hydrocarbons.

I was fairly proud of my newly landscaped back alley. It was fairly level. With the consistency of yogurt and granola, my oozing gravel sludge pool looked perfectly fine from a distance. While the old-growth Chinese cooking oil had a distinctly foul smell, it did not have the vomit-inducing sharpness of this next job.

At the time, my dad managed at a company that offered rental and cleaning services for things like uniforms, aprons, floor mats and kitchen towels. He became aware of a small gap in their regular operations. Occasionally a truck would return with bags of soiled linens that, for some reason, were not entered directly into the sorting and cleaning process. He would let me know and I would put in a couple hours at their sorting facility after school.

It was immediately clear why these deliveries couldn't be processed normally. They were special. The linens in cloth sacks about the size of a regular black garbage bag would normally weigh twenty-five to thirty pounds. These ones, however, suspiciously weighed almost twice that.

The bags also didn't feel soft and fluffy like you would imagine a sack of towels should feel.

I would lug a bag up to the sorting table, empty it out, and discover that, yes, the added weight was, indeed, biomass. Each bag contained its own thriving ecosystem. The rags were packed and glued together with grease, syrup, sauce and gravy, and would pull apart like a sticky fig sussed out of a fig cake. Inside the clump of cheesy rags was a hot core crawling with life – handfuls of maggots or longer, more robust-looking larvae, feeding on every type of restaurant debris possible, always including at least a few chicken wings.

The smell was sharp and mouldy. It smelled dangerous.

My gag reflex was on high alert as I pulled all the rags apart, separating them into different bins according to their size and type. Anything non-linen, and there was a lot, would fall through the grate of the sorting table. Sometimes writhing lumps of last week's lunch had to be forced through the grate.

Yet, even low-pay biohazard sanitation work can be fulfilling if you're in the company of great people. My girlfriend, soon-to-be fiancé, out of pure love and devotion, chose to work alongside me on many of those days. Fitting that the money from this job would pay for the first installment of Rose's engagement ring.

The week before our wedding, in the August heat, I found myself in a sensationally filthy barn renovation.

It was a duck barn with raised, grated floors. The grates were on two-foot concrete pony walls that created four-foot-wide channels the length of the barn – probably about ten channels of one hundred and twenty feet. At one end there was a series of large tanks feeding a water header. The tank valves would be opened quickly and the rush of water would flow down the channels, flushing all the duck poop into a sump at the far end.

At least, that's how it was supposed to work. The system had been in service for years but had never really worked. It turns out duck feed actually includes small amounts of sand which help the ducks digest the feed. And the sand, not soupy and snotty like duck poo, did not flow down the channels very well. A few pieces of sand would get stuck and slow the flow. A little sandbar would form and eventually grow into a total blockage of the channel, blocking the sand, the poo, the spilled feed and all the dust and little feathers that fall off of 10,000 ducks. The stagnant water helped bind all contributors into a homogeneous mass.

"Tear it all apart and clean it up, boys," were the instructions.

We would eventually totally redesign and rebuild the barn floor, but first we needed to expose the original concrete and make it spotless.

Teenage boys love a challenge and so, with gusto, five of us, all former classmates and friends, tore into the old flooring. We attacked the floor panels with competitive enthusiasm, swinging breaker bars, crowbars and framing hammers. Spread out through the barn, all of us were yelling back and forth, volunteering a play-by-play of our demolition conquest.

The heaving and smashing and yelling and laughing escalated into hoots and hollers as we each removed sections of the grates and began breaking through into ten inches of effluent. Teenage boys know how to appreciate multi-layers of disgustingness more than most, and we were about to get our fill.

How we didn't notice beforehand I'm not sure, but all that spilled duck food fermenting in the stagnant water had attracted a critical mass of mice. Safe in their little sewer for generations, hordes of mice had never prepared for this moment, and neither had we. So, as we entered the holy land of Rodentia, it got crazy. Mice were climbing up our boots and coveralls, mice were running away from us and towards us, mice were simply everywhere. At first stunned, it took only a few seconds to turn this mayhem into a positive.

"Hey Jeff!" and I turned just as a mouse sailed by.

Enlightened, I grabbed the nearest mouse by the tail and flung it at my nearest coworker.

"Heads up!"

It was a mouse fight.

Splashing saucy waves of effluent with every movement, we all scrambled to grab mice and hurl them dodgeball-style around the barn. Already euphoric, it got better the second we remembered the shovels. If someone is going to set me up with a nicely lobbed mouse, why not hit a home run? And so we dove for mice, threw the mice, and we hit the mice out of the park. It was a free-for-all. It was like every boy's dream. And by the end of the day, we were covered head to toe with duck poop, fermented food and mouse guts.

The next day was way lamer with most of the mice gone. But it was still pretty gross as we sweated a long day shoveling out all the sludgy waste. By Wednesday, I had spent three days in that barn. I went home and showered, and when I came out, in a flash of self-awareness, realized one shower didn't cut it. I showered a second time – to no effect. I was getting married on Saturday and realized that bathing in beaver castor for the rest of the week was not an option. I would have to quit a couple days early. Thankfully, three days was just enough time to shed the lingering aroma.

This job, like a lot of dirty jobs, was good. I got paid, I felt productive, and I enjoyed the company of my coworkers. I wouldn't want to work these jobs forever, but they served a great purpose. It would be another decade before I fully figured it out but for now, at least, I knew work was not the enemy.

Rose and I were married two weeks after my nineteenth birthday. None of our friends or family ever tried to dissuade us, but I'm certain there was a little nervous chatter behind closed doors. I'm sure my parents had to make excuses on my behalf.

"Oh, you know Jeffrey, if we tried to talk him out of it, it would only encourage him," or comments of that nature.

The way I saw it, I really had no choice. When you're handed a winning lottery ticket, what else can you do but cash it in? And I couldn't imagine why I'd want to pursue life by myself when I could do it with the company of my best friend. Not once have I second-guessed getting married young. Having essentially grown up and matured together, we are truly a strong rope of two strands.

One week later, I was in tech school at the British Columbia Institute of Technology (BCIT), and, within days, I knew it was a good choice. We were actually designing and building things.

Less than six months after she said "Yes!" we were executing the first stage of our plan. Rose had completed her legal office assistant course and was making a respectable twelve dollars an hour – almost five dollars more than minimum wage and three dollars an hour more than I had been making in construction. With a little bit of savings, some cash gifts from our wedding, and a little help from our parents, we had paid for my school tuition and rented the basement suite without borrowing any money. Aside from work I could get in the summer, Rose would be the sole breadwinner for the next two years. Once I graduated, I planned to get an even better-paying job. We would quickly save up $25,000, buy some land in the woods way up north, build a cabin, and, boom, we'd have it made.

I guess at nineteen it's possible to have some major blind spots. But they would become clear to us soon enough.

Brand New Baby and Ancient Technology

B CIT was the first formal education I actually enjoyed. It was such a wild concept to learn things with actual purpose. Instead of reading and regurgitating information, throwing it away and moving on, we were learning math concepts, principles of physics, and manufacturing fundamentals to design and build stuff.

I believe a fundamental human need is to be productive. You could argue that there is some amount of productivity in schoolwork – but your argument would be as flimsy as a grade-nine noodle-bridge science experiment. By far, most school assignments are make-work projects. If the content was truly useful, it would be easy to respect students' time and guide them in the production of something functional. I have a beef with the state of education too big to sort out here, but let me say that after a less-than-inspiring high school experience and a disappointing semester in university college, tech school greatly exceeded my expectations.

Within a matter of weeks, we had designed, fabricated, assembled, powered up and tested our own DC power supply. I had built a product – ten times more productive than a whole semester in my Associate of Science degree program.

It was intellectually very rewarding but also a seismic shift in lifestyle. Just weeks before, I'd been labouring physically for fifty hours a week. I played sports and was an avid outdoorsman – hiking, hunting, and fishing whenever I could. Now I was commuting two to three hours a day, in class eight hours a day, and doing homework evenings and weekends. On top of this, anytime I came home, my fantastic new wife would stuff me full of food. From August to December, I morphed from a lean, one-hundred-sixty-five pounds into a semi-rotund two hundred pounds. Although some people mistook my weight gain for lack of activity, I knew that I was just swelling up with new knowledge. Rose never seemed to care, or at least never said anything. She was quietly satisfied that, as a rookie wife, she was clearly capable of keeping her husband fed.

Rose and I ran our freedom plan like a well-oiled machine. I was crushing it at school, competing with the best in the class. Rose was putting food on the table financially and literally. And now, midwinter, it was time to look for a summer job. I was hoping to explore the north, so we travelled to Fort St. John to look for work. It was a bit of a long shot looking for technical work, with only a single semester under my belt, but that was my goal.

Uncle Dave, who was living in the tiny town of Hudson's Hope, put me in contact with a guy from his church. Martin was a manager at BC Hydro's largest generating station, GM Shrum, at the WAC Bennett Dam. I didn't realize it at the time, but it was difficult for BC Hydro to attract engineers and technologists out of the cities to a tiny, remote town at the 55th latitude. They offered me a summer job working in the Protection and Control department (P&C).

I had no idea what that was, but I was thrilled to get a high-paying job in my field of study and in the most beautiful corner of the most beautiful province in Canada. I was thrilled, but Rose had more of an I-think-I'm-excited kind of response.

Rose and I had both lived our entire lives exclusively in the Lower Mainland of British Columbia. I grew up in the same house till I got married. Rose's family had moved around a bit but mostly stayed in the Lower Mainland area. I had seen a few different areas of the province. My grandparents had lived in Midway in the Boundary District. Aunt Joy had taken us hiking in Manning Park, the Stein Valley and the West Coast Trail, and Uncle Dave took us hunting in the Peace area. I'd been on fishing trips to the Thompson-Okanagan and even into Charlotte Lake in the Cariboo. But Rose had hardly been anywhere until she met me.

With all my talk of moving to the woods and getting property, this was finally getting real. It was just for the summer, but it was still 1,200 kilometres north. She agreed to the move, but I knew she was uncertain. My certainty made up for her lack of it. I knew how beautiful it was up there, and I knew she would love it. Plus, we had family there – Uncle Dave, Auntie Wendy, and our little cousins, Jenna and Kristy.

That summer we undertook our first big adventure together, moving north and exploring the wilderness. We saw spectacular northern lights, caught our first northern pike, hunted and harvested our first black bear together, fixed our truck when it broke down and walked for hours through sweet-smelling alfalfa fields in the forever sunsets of the north.

Dave and Wendy gave us a sweetheart deal to stay in their basement. I would help Dave around their lovely, ten-acre property in exchange for room and board. We hunted the early elk/moose season, unsuccessfully, and just as the fall season was opening, we had to leave. The whole magical season left us wanting more. It was certain we would be back.

Work at the dam was fascinating. The amount of money, expertise, and coordination required to maintain, repair, and upgrade was eye-opening. With eighty-five full-time staff, an earth-filled dam a mile across and five-hundred-and-fifty feet tall, the largest man-made lake in North America, and 2700 Megawatts (MW) of electrical power output, it was hard not to feel like I was part of something grand.

I recall my first tour of the power station, and my overwhelming and recurring thought was, "How is this equipment so old?" I was going to tech school so I could build rocket ships and maybe work for NASA. I was going to program microcontrollers and design tiny robots. And yet the electromechanical protection and control devices at the dam were huge, clunky beasts originally designed and built in the 1940s, and they hadn't changed.

Protection and Control referred to the protection and control of utility-scale electrical devices. Turns out 300 MW generators and transformers are very expensive. Their cost more than justifies highly-specialized devices to control them within design criteria and protect them from damaging faults. They have multiple flavours of short-circuit protection, open-circuit protection, overspeed protection, etc. Anything that can go wrong on a utility power system must be identified within milliseconds and action taken to isolate the problem.

A Protection and Control technologist specializes in the installation, testing, commissioning, troubleshooting, and maintenance of these devices. A tiny niche in the world of electrical engineering technology, they play an outsized role in the operation of the grid.

The physical scale of the dam and its reservoir, and the associated power equipment, is impressive. Twenty-foot diameter penstocks descend hundreds of feet below the intake and into a subterranean powerhouse, where they meet the turbine. They wrap around like a snail shell, forcing the water through the wicket gates and into the turbine blades. The turbine spins a forty-inch diameter steel shaft connected to the generator rotor. During a shutdown, multiple crews could be in the scroll case, the turbine room, and inside the rotor. If you were playing a game of sardines, you could probably pack fifty people inside the rotor itself.

Very cool, but I wasn't one hundred percent sold. The equipment was old and dusty, and I wanted to work on something leading edge. Most of the equipment was older than my dad. I wanted to invent things, and there was no inventing here. At the same time, there was so much to learn – like how exactly does a spinning magnet create electricity and, more nuanced, how does that freshly generated electricity apply real force back to the rotating magnetic field and restrict its velocity? Now that I thought about it, I definitely had to figure these things out. But did I really want a job playing with a bunch of old stuff? Shouldn't I be inventing new stuff?

Back at school in the fall, I changed my second-year option to a dual diploma in both Computer Control and Power and Industrial Electronics. The Computer Control option was more interesting, but the Industrial Power option was the prerequisite for a P&C technologist at BC Hydro. Consumed again by school, and with next year's career choices coming fast, we were about to get some firm direction.

We were driving to Rose's work Christmas party in downtown Vancouver. It was dark and rainy. A few blocks before we arrived, Rose asked if we could pull over.

"Ok. . . ?"

In the McDonald's parking lot on Main Street across from Science World, with homeless people milling about, she looked at me and teared up. She was not normally overly emotional, and in fifteen months of marriage, I don't think she'd ever insisted on a heart-to-heart or whatever this was about to become.

"You know how the pill wasn't making me feel very good?"

"Uh huh," I replied with a nod.

She looked at me intently and, after seeing no signs of intelligent life, continued.

"Well, do you remember how we talked about what might happen if I went off?"

"Yes," I said, pleased that I seemed to be doing so well on this little quiz.

"Well, I'm pregnant," she blurted out, and started crying.

I was still putting together the flowchart: Pill made you feel bad -> Go off the pill-> Pregnant.

"Really!?" I finally offered.

"Yes," she replied, sounding apologetic.

I was still processing the implications. That meant Rose would become a mom... and I would be a dad.

"Are you ok? Is that ok?" she implored.

I had a healthy grasp of biology, but I had never really thought about Rose having a baby or me being a father. I mean, in a science textbook it makes perfect sense, but in real life, that would turn me into an entirely different person. I would transform from a kid and a student and a charming young husband into a father. And fathers were old. No one had ever spelled that part of it out for me.

Either way, Rose appeared overwhelmed with uncertainty, so I gave her a hug.

"Well, that's great! Of course it's OK!"

"Really!? Are you sure?" she asked, needing more affirmation. "Are you happy?"

I had never thought about it, but a pregnant woman without a committed husband must feel very vulnerable. Finally, sensing the painfully obvious, I offered my wholehearted affirmation.

"Yes, it's OK. It's terrific."

A look of relief, then joy and then excitement overcame her.

She leaned over and hugged me again with much more conviction than normal and said, "I'm gonna be a mom!"

That's all Rose ever wanted. She wanted to be a mom. I clearly never understood that – I'd probably never asked her. And Rose doesn't verbalize all her deepest desires; she needs to be carefully observed to discern her thoughts and feelings. Of course, no one ever accused me of being too sensitive, which is why I'm still learning new things about Rose twenty-five years later.

Now, on the way to the Christmas party, we had a lot more clarity for the life and career decisions to be made in the next few months. Rose wanted to be a stay-at-home mom, and I wanted that, too. At nineteen, we didn't let a lot of facts get in the way of our happy young marriage. Now, at just twenty, we were starting to fill in the details.

We did some quick math. My school would complete in June and Rose would probably stop working in July. I could probably work June till December, then be back to school January to June. Rose would not be working during that time. She would be at home with the baby. Yep, no problem, things would get pretty skinny during next year's spring semester, but we would get by.

Sarah was born on my twenty-first birthday – what a fantastic present. A first kid can be very stressful, especially on a minimal income, but that was not the case for us. I had no idea what I was doing, but Rose was made for this and had been programmed with all the code for mothering. She unflinchingly woke up at all hours of the night. She fed and changed the baby and did most of the housework with Sarah on her hip. I always knew Rose was a winning pick, but to see her elevate from first-round wife to first-overall mother made me proud.

I took on some contract work building small, temperature-alarming devices for the duck barns that I had worked at a few years prior. It allowed me to work mostly from home so Rose, Sarah and I spent most of our time together for the first four months.

We survived my last semester at school thanks to my parents, who let us stay rent-free in their basement suite. I emerged from school full of knowledge, plans and ideas and approximately $0 dollars to fund them. It was time to find the land of opportunity.

Headed North

I had put out resumes to a few different tech companies in the Greater Vancouver area – from cool little startups developing tiny electronics to a giant semiconductor manufacturer. I remember applying at PMC Control, which designed, built, and installed navigation and steering systems for large ships such as freighters and navy vessels. They were attractive high-tech jobs, but were all in the city. I had also applied back to BC Hydro, and they had positions all over the province.

With a new baby, it made a previously difficult career decision into a much easier lifestyle decision. I could have a cool job and our little family would grow up in an apartment on a busy street in a smoggy city or I could take a boring job and take my family into the outdoors, somewhere in the great white north of BC – an area larger than Germany but with the population of Greenland. Mathematically, the northern area is home to less than one person per square kilometre, versus over 5,700 people per square kilometre in downtown Vancouver.

I accepted a position with BC Hydro as a P&C technologist in training. In order to get experience in generation, transmission and distribution, they required me to work in two or more regions. My first rotation was Prince George. This small city is basically in the geographic centre of British Columbia – a nine-hour drive north from Vancouver but still a ten-hour drive south of the Northwest Territory border. Going west, it's an eight-hour drive to Prince Rupert and the Pacific Ocean while east is a four-hour drive to the mountains of Jasper on the Alberta border. The city is in the middle of the boreal forest at the confluence of two great rivers, the Fraser and the Nechako. It's not really a destination. It has a lot of good blue-collar jobs in forestry, lumber and paper mills, railway, mining, transport and construction, but it lacks the flashy sea-to-sky vistas of Vancouver and the wineries and orchards of the Okanagan.

We had only ever driven through Prince George and weren't particularly impressed. This was not our first choice, but it was where we were going to be for the next one or two years. Little did we know the city would play into our plans long into the future.

I started my new job in June of 2002 and my pay was a healthy nineteen dollars and fifty cents an hour. The whole baby incident delayed our freedom plan a bit and this temporary rotation in Prince George would delay it some more, but earning almost forty grand a year would keep us inching toward the goal.

We upgraded our two-wheel-drive pickup to a four-wheel-drive Jeep Cherokee. We found an abomination of a canoe with a "For Sale" sign leaning up against someone's fence, and a hundred and fifty dollars later, it was strapped to the roof. A patchwork of peeled paint only partially covered up what must've been a drunken attempt at fibreglass repair. The gunwales were cracked and the homemade plywood seats were just screwed through the side of the boat. It was awesome.

We were like redneck rock stars with our Jeep and our canoe and our little baby bouncing around in the backseat.

We spent that first summer exploring back roads and catching fish in obscure lakes. We went to Mackenzie Lake a lot, which was just half an hour from our newly rented house. We spent more than one bluebird afternoon paddling on those still waters and always seemed to catch a couple trout by sunset. Rose and I would take turns paddling, fishing and stopping Sarah from falling in the water. Eventually, we discovered a little bucket of rocks could keep her entertained for at least half an hour as she threw each of them, in succession, into the water. Sarah became a fisher for life.

Rose and Sarah and a Rainbow Trout

Most of our adventuring was on the weekends, so Rose and Sarah were home a lot by themselves and we were eight hundred kilometres away from friends and family. We had a lot of family and close friends back in the Lower Mainland, and there were times when we both felt a little out of sorts. And for that, I am very grateful. Unless forced to make our own way, it would've been easy to drift along, leaning on family for support and friends for entertainment. Rose and I just had each other, and together we had to make new friends and support each other while developing new interests, skills, and goals.

Adventurers, explorers, and leaders need the freedom to forge a new way. And a familiar, comfortable life is hard to pull away from. Because we were building a new life in a new place, it was much easier to try new things.

Of course, we made new friends – fantastic families we would have never known had we not uprooted ourselves. And we continued to aim for our Freedom 25 plan. Now, with a little more stability and only three years away from the target, it was time to take a close look at the numbers. Before we could write anything up, Rose announced, yes, another baby. Our eighteen month stay in Prince George was capped off with the birth of Abigail.

Sarah's birth was a long, drawn-out, painful, impersonal, and clinical event in a stereotypical hospital full of random strangers in scrubs. Wires, hoses, needles, and beeping machinery set the tone.

Abigail's birth was the polar opposite. She was born quickly, without medical interventions and with the encouragement and support of a midwife and doula in our own home. Within moments of birth, Rose and the new baby were snuggling in our own bed. For that experience alone, Prince George earned a warm spot in our hearts.

Our next and, hopefully final, rotation was back to Hudson's Hope. We were finally in God's country. Unfortunately, Uncle Dave, Aunt Wendy and the girls had recently moved to Fort St. John, so we were again in a new community with no friends or family. While we had been here for a summer, on our previous stay we hadn't really had to fend for ourselves or establish ourselves in any way. Now, with the prospect of being here for some time, our perspective was different. We would need to make new friends again and find our place in the community. There were lots of young families, which was a great surprise – other young moms for Rose to relate to – and a small, but welcoming, church congregation that made us feel at home.

Working back at the dam was familiar but felt different. A few people on the crew had changed in the last three years, and that changed the dynamic a little bit. But mostly it was that I was now a P&C technologist, not just a summer student. They were chronically short-staffed, so I was thrown right into the thick of it. There was no time for mentoring. They gave me the maintenance instructions, showed me the technical library (an actual wall of books), and let me loose. It was figure-it-out-as-you-go mode. Not a very efficient way to get the work done, but learning the hard way definitely contributed to developing a deep understanding.

The dam and adjacent generating station was a perfect movie set for *Mission Impossible*. Hundreds of feet below the surface of the ground, the cavernous powerhouse was about fifty metres wide and two hundred and seventy metres long (about three football fields!) with arched walls and ceiling of exposed rock, ribbed with concrete buttresses. Giant hooks, thick as oak trees, dangled from five-hundred-tonne overhead cranes, as if fishing for prehistoric monsters. The thundering rush of over 2,000 metric tonnes of water per second and 2,500 megawatts of rotating machines properly sounded like an advancing and imminent natural disaster.

The powerhouse floor, aside from just a few freestanding electrical cabinets scattered along one side, was flat and smooth. It was mostly concrete but had two hundred square metres of huge removable steel panels above each generator. On rare occasions, the powerhouse was empty and clean, like an abandoned air force hangar. I could imagine an old B-52 sitting in disrepair at the far end. But, most of the time, it was full of gigantic electrical and mechanical equipment.

During annual maintenance, the 250 MW machines were disemboweled. After crews unbolted the necessary connections, the overhead cranes were linked together, giving them the required strength to pull them out piece by piece. It left a three-story hole in the powerhouse floor, various dangling connections hinting at the complex systems concealed below.

Taking a narrow stairway down to the generator floor, a seemingly endless hallway lined with pipes and electrical wiring faded into the distance. Like a tiny explorer inside a mountain-sized C-3PO, the variety and quantity of infrastructure was overwhelming. Massive concrete walls caged the giant spinning beasts and deadened, but did not eliminate, their awful howls. Banks of ten-foot steel panels were filled with every manner of electrical device. They looked benign but actually controlled every aspect of the power plant. Many were faded relics with little glass windows exposing a microcosm of windings, magnets and springs. More like toaster-sized dioramas of industrial artwork, it was forgivable to underestimate their importance.

These P&C devices could detect, in just a couple milliseconds, almost all possible electrical and mechanical faults. They could also control the machine to maintain desired performance. The power station, like the human body, has a lot of stuff that's nice to have, but only a few that are "life and death." When catastrophic failure is imminent, nothing matters except a couple of tiny wires and the electrical signals they carry. A 20ms blip of voltage can shut everything down; stop the water, stop the machine and stop the electricity. It may look like a phone charging cord, but unplug it and an entire city instantly goes black.

It is common for fallen trees in a windstorm to cause power outages to dozens or hundreds of customers. But a protection operation (or worse, an accidental operation due to personal error) at a giant facility like this could be felt in another country and cost millions of dollars, as I would eventually discover.

Another set of stairs led down to the turbine floor. Strangled with piping – like the roots from some massive fig tree – it immediately felt more dangerous, with warning and hazard signs hung on every door, portal and panel. This floor housed the large hydraulic pressure unit that actuated twenty-inch-diameter hydraulic pistons. These, in turn, connected to the wicket gates and precisely controlled the amount of water to the turbine.

A thick insulated steel door and then a short concrete hallway led into the octagonal turbine chamber. It was necessary to wear earplugs underneath my earmuffs. From the little concrete hallway, I could step out onto a grated catwalk that circled the chamber and put my feet just inches above the hydraulic workings. Any body part that found its way down there would be neatly popped off. It was the perfect place to tussle with a villain.

In the middle of this screaming, grinding, oily torture chamber was the giant, shiny, stainless shaft that connected the turbine below to the rotor above. As I gently reached out and resisted the rotation with the back of my hand, I imagined needing the additional power of 350,000 horses to stop it.

The next floor below was undesirable real estate, mostly vacant of equipment, and the bit of cabling and instruments here and there seemed neglected. It was distinctly colder and wetter, with condensation on the walls and water dripping and trickling from spouts and pipe stubs into gutters along the walls. Small concrete passages meandered to various installations and, sometimes, to nothing at all. One in particular led to what seemed to be a submarine hatch. It was a circular plate about two-and-a-half feet in diameter and at least an inch thick. It had two simple steel handles, one on each side, and was secured with one-inch bolts spaced so closely you could hardly get a wrench on it.

The hatch was centred in a bit of a truth window, a break in the concrete exposing a curved steel surface below. It was the scroll case. Behind that hatch was over 200 PSI of ice-cold water from the lake 150 metres above us.

From this level the only thing below was the sump, the lowest point in the powerhouse. It was the kind of place you might expect to find Gollum. I could climb down a rusty steel ladder through various shafts leading straight down into a watery pit. There were also a few tunnels that seemingly led to nowhere. I remember exploring the labyrinth of passages and finding one that led to an open cavern in the rock. It was a large cave, for some reason never utilized and never closed off – an artifact from construction days. In these remote corners, it was common to find a section of cable tray void of any actual cables but curiously hosting a blanket and pillow. Weird. There was ample space, and the shadowy catacombs almost invited nefarious activities. It seemed necessary to keep peeking over your shoulder and repeatedly scanning every corner.

The whole facility was half magnificent technical marvel and half old dirty mess. Regardless, it was, by definition, a powerful place, and I was about to learn just how far its influence reached.

One of my first assignments was to perform preventive maintenance on the electronic governor. Kind of like the cruise control in your car that pushes and releases the gas pedal, the governor controls opening and closing of the wicket gates, allowing more or less water to the turbine.

A bit like a patchwork quilt, the governor control system had been revised, modified, tweaked and upgraded, but not always captured in the schematics and documentation. Some things were not documented at all – they were just common knowledge to everyone except the new guy. As I studied the maintenance standard I quickly ran into a roadblock. One of the very first steps before testing was to "isolate the governor output." I couldn't seem to find an isolating switch. After asking another tech, I learned the output switch was down on the turbine floor. I was working on Generator 7 (G7), but the stairwells leading up and down were only on every second generator. I walked over to G8 and marched down to the turbine floor. Looking around at the mess of pipes and valves, it was all a little disorienting. I looked across the hallway, and there it was – the governor control cabinet. I walked around and found the door. It had a giant, red number EIGHT on it, about three-feet tall and placed directly at eye level. I did not see it.

I opened the door to discover all manner of small copper pipes, hoses, manifolds, valves, and machined parts – it all looked suitable for a 1000 PSI hydraulic system. Thankfully, there wasn't much for wiring, and so there was only one set of knife switches that caught my eye. Just in my second year as a P&C tech, I had already adopted a very careful manner. I had heard all sorts of stories about how one switch or one wire could cause a regional blackout. It was part of our training to confirm in every possible way that we operate the correct device. In this case, there was no label on the switch, and there was no drawing showing where the switch was. Its location was just "common" knowledge.

I stared at it for quite a while.

At some point, I just needed to get something done. And so I reached out and pulled the switch.

Instantly, I heard a large groaning sound of machinery and the corresponding rush of water. I was alarmed, but in the bowels of such a powerful and active beast, I couldn't take every scary sound as a sign of imminent failure. I ran back up the stairs and over to Unit Seven, eager to finally get into some actual testing.

About fifteen minutes had passed when the duty electrician stopped by.

"You guys haven't been doing any work on G8 have you? The metering seems to have totally failed, it's swinging from -150MW to +250MW."

I reflexively replied, "No, I'm just working on G7."

We stared at each other.

In that moment of silence, we could both hear an accelerating rush of water coming from the adjacent unit. That violent sound, like whitewater rapids, gripped my imagination and instantly weighed like a stone in my stomach.

So. Much. Power.

It seemed out of control – almost like it had no control…

The electrician continued on his way to G8, and I ran after him. Down the stairs and over to the governor control cabinet. I beat him there. This time, opening the cabinet door, I saw it. The giant red number EIGHT. I wasn't sure what would happen, and again I hesitated. Was this like a stab wound where you leave the knife in to slow the bleeding? I didn't know. But I had to do something, so I turned my head to the side, closed my eyes, and shoved the switch back in.

Another monumental surge of water as the wicket gates sprang into action, and then total calm.

My boss smirked as I shamefully recounted what had happened.

After letting me sweat through an awkward explanation, he said, "Well, luckily it's not actually that big a deal because the unit didn't actually trip offline. It's not technically considered a personnel outage."

I breathed a sigh of relief.

"But the frequency disturbance that it caused was registered in California," he said as he handed me an event record printout from San Diego Gas and Electric.

"Ok..." I said and paused for his reply.

"You'll need to write a report describing the incident and its impact for the regulatory body. There won't be any record on your personal file," he assured me.

As I investigated further, it turned out my little experiment had exposed an unknown failure mode within the governor. There was supposed to be a small spring in the hydraulic valve that tends to close off hydraulic flow in the absence of an electrical control signal. This had never been tested before, and we had proved it didn't work. I referred the problem to the mechanical engineering department for corrective action.

These were complicated machines with complicated and overlapping systems of control. Thankfully, my little mishap did not land me in the MW club. My good friend and co-worker, however, was not so lucky.

He was performing maintenance testing on a breaker fail circuit. One of the highest-risk test procedures, it's not for the faint of heart. This particular circuit had the ability to trip off major parts of the generating station. When other protective systems failed and multi-million dollar equipment was on fire, the breaker failure circuit would react and trip off almost everything.

Imagine if your microwave started arcing and sparking. Normally, a circuit breaker in your house would automatically shut it off. But what if the breaker didn't trip and it just kept arcing? Your next hope would be that the main breaker would trip and shut off power to your whole house. It's better for your house to lose power than to burn down.

In a similar way, the breaker fail circuits on the power grid are capable of de-energizing large areas to protect against catastrophic equipment failure (aka huge electrical fires). Like all important protection and control circuits, they are tested at regular intervals to ensure correct operation.

Testing requires a very specific sequence of operations to confirm that it works properly but avoids actually tripping anything:

- Disconnect certain trip outputs by opening a whole bunch of switches
- Ensure only one particular switch is closed
- Energize a specific relay
- Check that a certain other relay operated
- Put everything back
- Repeat, but with a different combination of switch positions

The switches to be operated are hidden among thousands of others, in different panels, different rooms and even different buildings. Pre-approval is required from the plant operator and is recorded in the station log. It's no joke. It is, however, statistically probable that if you perform this task enough, you'll eventually slip up. All it takes is one switch to be in the wrong position. This was exactly the case when my friend joined the GIGAWATT club (a gigawatt is one billion watts -1,000,000,000).

With all the isolation switches pulled (opened), he placed one critical switch in the 'on' position, double-checked everything, and gently tapped the relay contacts together. No one could have imagined the cascading effect that would follow.

The post-analysis took days of sorting through countless sequence-of-event records and digital fault recorder waveforms at multiple stations across the province. It also required significant consultation with our central engineering and planning group. It all started with a tiny slip-up and the wrong switch. Instead of testing a carefully blocked output, it tripped off two generators at maximum power, cutting power output from 550MW to zero in approximately .08 seconds (80ms). 550MW is equivalent to the average power use of about 550,000 homes.

In short, it shook the power system.

The significant power loss triggered an old, nearly-forgotten power controller, which sprang to life like Homer Simpson and incorrectly tripped another 200MW. How that controller existed for decades with this internally flawed logic was anyone's guess but, regardless, the total lost power was now 750MW.

200ms after the relay contacts sparked closed, the power swing swept across the province and triggered another mis-operation at a BC Hydro power station one thousand kilometres away. Mica Generating Station, on the Columbia River system, was pumping out about 1000MW at the time. A five-pound hunk of electronics there sensed the system disturbance and erroneously tripped the entire station.

Less than a second after it started, the whole western electrical grid of North America jarred with the loss of 1.75 gigawatts of power.

It took months to settle the investigation, fines, and intra-utility power charges, but only an instant for my friend to reach legendary status.

Fire and Ice

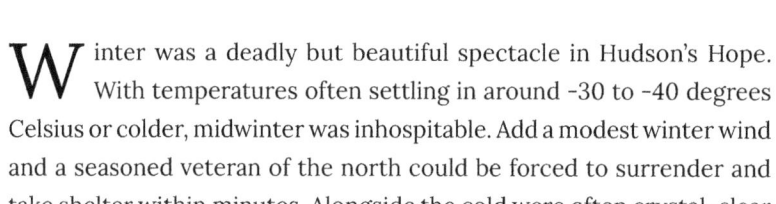

Winter was a deadly but beautiful spectacle in Hudson's Hope. With temperatures often settling in around -30 to -40 degrees Celsius or colder, midwinter was inhospitable. Add a modest winter wind and a seasoned veteran of the north could be forced to surrender and take shelter within minutes. Alongside the cold were often crystal-clear blue skies, powdery white snow, and a few, short hours of glorious sunshine. The beauty, purity and severity were always tempting, offering an icy white face as a badge of bravery for all who dared venture out.

The winter forest was frozen and quiet. But if I sat still, especially on days where the temperature had just plunged, the silence was punctuated by trees exploding in the distance – like the crack of a gun or a thunderclap. In just the right conditions, it seemed the entire forest might succumb as one tree after another ruptured suddenly from the pressure increase of its frozen sapwood. More constant was the plethora of sounds the ravens made, "Gloock, Gloock?" their calls floating through the trees, asking a tireless stream of rhetorical questions.

All the mud disappeared, and previously inaccessible muskeg swamps became fair game. If I could bear the weather, the entire wilderness was mine to behold. With proper clothing, I could work outside all day at -20°C. Internal biological thermometers are calibrated differently but, for me, I know it's below -20°C if my nose gets little ice crystals inside it with each inhale. I enjoy being outside at -30°C, but must work continuously to not get chilled. Regular, steady breaths are imperative to minimize the stabbing cold in your throat. Temperatures at or below -40°C can be enjoyable as well, but for a different reason. It's somehow very satisfying to be outside at that temperature and know that you are, at least for the moment, defying death. If it's windy, all bets are off.

We were in the north for all the seasons, eager to experience and conquer each one. One gorgeous Saturday in January, Rose and I strapped three-month-old Abigail and two-year-old Sarah into their car seats for a midwinter adventure. In Prince George, I had souped up our 1990 Jeep Cherokee with a three-inch lift kit, thirty-three-inch tires, and a Lock-Right automatic locker for the rear Dana 35. It was virtually unstoppable and I was going to take my young family exploring new territory in the dead of winter to demonstrate it.

I had been down the Gething Mine Road years before with Uncle Dave and Auntie Wendy, but never drove there myself – and not in the winter. When we got there, it was clear no one had driven it for some time and it was blanketed with a foot and a half of undisturbed snow. The Jeep bounced through it, cutting crisp, deep tracks in the snow and kicking up a cloud of tiny crystals. With a bit more throttle, the snow flew past the windows and occasionally blew over the hood like a schooner in the high seas. It was like we were filming a Jeep commercial. Our jacked-up Cherokee wove through the forested winter wonderland, powering a huge white plume into the golden-hour sunlight. At just two o'clock in the afternoon, the sun was already low.

And that's when the Jeep stopped working. It hiccupped a few times. Rose and I locked eyes. And it died.

It was a punch in the gut.

"Why did it do that?" Rose asked tentatively and innocently.

"I'm not sure. Maybe we should turn around now," I suggested with delayed wisdom, downplaying my own concern.

I tried to restart it. The Jeep fired back up and we both let out a sigh of relief. As I backed off the narrow track into the ditch to turn it around, it died again. Rose shot me a momma bear look that threatened violence if I did not immediately fix the problem. I turned the key, and it coughed and choked and then ran with all the regularity of an amateur rap battle.

Then it sputtered into silence again.

Rose was visibly anxious and I intentionally avoided eye contact. To assure her this was some minor issue, I did the man thing and decided to have a peek under the hood. I threw open the door and was met with a shocking wave of arctic air. Sitting in the heated cabin, it was easy to forget about the harsh conditions we were in. As I felt it flood the Jeep, it hit me that, unlike just getting stuck or breaking an axle, a dead engine would leave us stranded without any heat.

The crisp cold air, at -25°C, no longer felt exciting and invigorating. It was now a deadly threat to our lives. I hopped out quickly, closed the door and popped the hood. I'm no mechanic, but I knew where the air filter was, and it looked fine. I had a look at the spark plug wires. They were all there. There were no obvious disconnected hoses or electrical connectors dangling around. Within a couple minutes, after poking through and handling a few metal parts, my bare fingers ached from the cold.

I hopped back in the Jeep, praying I had somehow fixed something, and turned the key. It cranked for a long time, sputtered, and died. Somehow the Jeep had lost its mind and totally forgotten how to fire all six of its cylinders.

A gorgeous sunset quickly faded and Rose and I went to work devising a plan.

"I could walk out," I suggested. "But it will take me all night. And who has a vehicle that could actually make it back in here?"

"No, that will take too long," Rose asserted.

"We could all walk out... But the girls... I'm not sure how we'd keep Abby warm," I debated.

Rose teared up at the thought. She grabbed Abby's tiny little hands and continued, "She could get frostbite... ."

"We could carry the kids inside our jackets, but they'd get wet with sweat, and we couldn't always keep their hands and faces covered," I stated, and, imagining our trek, I thought out loud. "I don't even know if we could carry them for that long, slogging through this snow."

Rose couldn't imagine a successful trek either.

"No, we have to stay here and stay warm," she replied.

We had planned to be driving for a couple hours and brought a few snacks. The dark and cold of night would last seventeen hours. We had brought three extra diapers and had already used one. We figured a wet diaper could get dangerously cold for baby Abigail. The Jeep would have been a valuable shelter from wind or rain in warmer temperatures, but it offered little for insulation. A little bit counterintuitively, we would have to go outside and build a fire. We rummaged through the emergency equipment in the trunk and evaluated its current usefulness.

Axe. Great!

Old canvas tent with broken poles. Maybe?

Tow rope? Probably useless.

Matches and a lighter. Perfect!

One sleeping bag. Could be a lifesaver!

"I'll build a fire," I offered.

I hopped out and set to work with urgency. Rose got some snacks out and put on a brave face for the girls. Abigail was fussy – something to do with getting a subzero diaper change, but perhaps more to do with the palpable anxiety of Mama. Like every good mother in a time like this, Rose only cared to protect her children. Changing Abby's diaper and feeling her chilled skin brought fresh tears. Quickly bundled back up, her little hands and cheeks still felt icy.

It was a bad decision to come out here alone. No one knew where we were, there were no cell phones and we were not prepared to spend any significant amount of time outside. Rose was distinctly aware of the risk we were in and was looking to me for leadership. I knew she was ready to do anything required; I was just hoping it wouldn't need to be anything drastic.

With plenty of dry spruce underbrush and our emergency lighter, starting a fire was not a problem. The axe made it easy to break off larger branches, and soon we had a roaring fire.

"Get out here, Rose! Bring the girls!" I called.

In these temperatures, we had to stand in the licking flames to have a net positive thermal influence on our bodies. Our hands and face absorbed the radiant heat, but the backs of our bodies had the heat drawn out of them by the devouring blackness of space. Not too challenging for adults, but how could we heat a baby by a fire? Could we hold them out kinda like thawing a frozen hotdog bun? Instead, Rose held Abby close and kept her mostly zipped under her jacket. She would alternatively let her head pop out to get fresh air and then tuck it back inside to keep warm. Knee high Sarah, under her own power, stood alongside us, next to the fire.

"Sarah, you can help me throw sticks on the fire?" I suggested.

She picked up a few burnt ends and threw them in the general direction. Rose watched her like a hawk. A foot closer and she'd fall right into the blaze. A foot further back and she'd start losing body heat. I cut a bunch of green spruce boughs, threw them down by the fire, and covered them with the canvas tent. We could sit on it or put the baby down without having to put her in the snow. Rose carefully distributed a small handful of fishy crackers, more to raise our spirits than for nutritional value.

Having made some progress, we came to grips with the challenge of somehow standing beside a fire all night long. It had been a couple hours, and yet it was only six o'clock in the evening. It was disheartening to consider another fourteen hours of darkness. For the next few hours, I collected wood and kept the fire going. It was a full-time job for Rose to keep Sarah awake and warm. She had fallen asleep inside the sleeping bag on the canvas heap, but Rose could feel that she was cold and thought it was best that she keep her eyes open. It's not very natural to keep your baby and toddler from sleeping, and not very enjoyable either.

I returned to the Jeep, suddenly thinking that maybe it just needed a timeout and, having learned its lesson, might fire up. It turned over, fired up just like normal, and ran smooth. Rose shot me a look, her face pure excitement.

"What?!" she exclaimed. "Jeff! Is it gonna run?"

The engine answered on my behalf with a sickly cough and haltering cessation.

"No, No, NNOOO!!" I cursed, and cranked the engine again for an abusive duration.

It had run properly for about 30 seconds but now stubbornly refused to start.

"What is wrong with you?!" I demanded, staring daggers at the dash. "You. Are going to kill us!"

Rose acted quickly to interrupt my spiraling tirade of frustration.

"Maybe if we just let it sit… Maybe it'll start again in a while?" she suggested.

It had been stone cold for a few hours and that seemed to give it a little life.

"I think it needs to be totally cold. We should try it again in thirty minutes," I concluded.

The next time I tried, I was ready. It fired up and I quickly shifted into drive and pulled back onto the road just as it died. We only moved ten metres closer to home, but it felt good to at least be out of the ditch and aimed in the right direction.

"That's better," Rose encouraged as I considered our strategy.

"I think we should wait longer next time, maybe it'll run longer," I concluded.

It was midnight the next time I tried to start the Jeep. We had spent ten hours facing the cold and were feeling exhausted. We struggled to balance the girls between freezing, burning and sobbing. We ate a bit of snow just to get some moisture. And we were out of diapers.

Discouraged about our prospects of hanging on to the fire, I proposed we pack up and get ready to move the next time it started. Maybe it would run for a minute or two and we could get a mile or two closer to home.

"Ok, Sarah," Rose started, "Let's clean up. We're gonna get ready to go."

We piled in and she gave me a brief smile, a façade of optimism.

"Ready?" I asked.

She hesitated and then replied "Yes," with forced resolve.

I fired it up and, wasting no time, gunned it. An instant rally race, I fought the wheel and Rose corralled our loose junk as it took flight with each bump. It was squirrely even on the straight sections and we drifted the corners. Unfortunately, as I had suspected, it only lasted for ninety seconds. We stuttered and lost power. The snow-laden road sucked at the Jeep tires like a giant spider web and we slowed to a stop. In frustration, I pushed the gas pedal hard to the floor, willing the engine to keep spinning. It backfired erratically but kept running. Unwilling to let it die, I maintained full throttle. The backfiring continued and then suddenly, like a rogue wave, an explosion like a gunshot shook our seats. BOOM! Rose grabbed my arm.

"Jeff! Turn it off!" she cried.

"No!" I shot back. The engine was running and I was going to keep it running.

It was a chaotic, shuttering cacophony. Even though I was in gear with the gas pedal pinned, we were not moving. Our shamble of a six-cylinder mustered the power of a lawnmower and, though the snow was fluffy on top, it was heavy and sugary at the bottom. In desperation, I threw it in neutral and pulled the transfer case into four low. Amazingly, we lurched forward and started crawling down the road. I presumed the engine was moving towards a catastrophic mechanical failure but, at this point, it didn't matter. We had been contemplating hypothermia, frostbite, and even death for hours – if the Jeep didn't survive, it didn't matter. The backfiring continued, and every minute or so it would build up and explode violently.

"Are you sure it's ok?" Rose asked.

I answered as honestly as I could, "No, but we have to keep moving."

I presumed we would get some heat but, clearly, with the engine barely running, it provided nothing. We were frigid, shivering and miserable. Without defrost, I was struggling to see the road. We settled in, silent passengers on a slow-motion, disaster-wagon trail ride. The constant backfiring gave off a rich exhaust smell that steadily grew stronger and, like frogs boiled slowly, we were soon engulfed in smoke.

Rose, alarmed, blurted out, "Smoke! What's burning?"

"It smells like burning plastic," I said.

As soon as we verbalized it, the smoke overpowered us. At the same moment, my eyes were drawn outside to the glowing orange snow alongside the Jeep.

"What in the ....??!!" I thought out loud.

Then all the alarm bells in my head started clanging at once. I threw open the door and hurled myself outside in one motion. Rolling as I hit the ground to look underneath, I saw the entire underbody on fire.

I frantically flung armloads of snow under the Jeep and against the sizzling frame. I wasn't sure exactly what was burning, but as the flames dissipated, I noticed the muffler. Glowing red-hot and appearing almost transparent, I understood what had happened. All that backfiring in the exhaust system had heated the steel until the undercoating reached its ignition point. I continued the full body effort to hurl snow at the fire, suddenly thankful for its abundance.

"What's happening?" Rose yelled, as she flung open the doors and jumped out.

"We were on fire, but it's out now," I offered, slumping to the ground in resignation.

"It's still burning somewhere inside!" she insisted.

I scraped myself up to investigate and rummaged through our cluttered gear, tracing the trail of sweet, toxic smoke. It was pouring out from the carpet right below Sarah's feet. I swung her from her car seat and quickly packed the carpet with snow. Scouring the entire jeep floor, I hit a few more smoldering hotspots.

Standing outside our smoldering Jeep in the still, black cold of night, we were right back at square one. The midwinter temperatures had threatened to kill us but the plentiful snow ended up saving our bacon. With all the windows open, I floored the gas and convinced the engine to run once more. Pulling back on the gas pedal just a little, I found I could keep it running, still with a terrible chop but without the massive backfires. Clearly, moderation was a lesson I was still learning. I popped it into gear, the wheels churned in place for a second and then slowly moved us forward.

"Get in! We're going!" I called to Rose, and she jumped in.

Even at just a few kilometres an hour, the arctic air blasting in through the open windows was hard to breathe but far better than the poisonous smoke still emanating from the carpet. In the wee hours of the morning, we edged the Jeep onto the paved road. It took another hour to get home, one eye always on the ground beside the Jeep for any signs of fire. The last couple of miles were downhill, and we coasted in.

Exhausted and humiliated, we were stunned by how close we had come to getting killed by nothing more than a cold winter's night. Feeling defeated, it was clear that despite my enthusiasm, I was still just a city boy. I had a long way to go to become the mountain man I intended to be.

Homesteads

Having worked two years for the lumber mill in Princeton, BC, my grandparents, still in their early twenties, made a big move. They had saved up enough money and figured they should get started on a little homestead. They toured around the Thompson-Okanagan region of BC, which was filled with small family farms and orchards. Grandad's parents had just immigrated from England and, together, they paid $5,000 cash for an old farmhouse and 93 acres in Notch Hill, near Salmon Arm. Notch Hill was not much of a community and was mostly named for the water tank refill station on the CN Railway that passed through. Years later, that railway would deliver an incredible surprise from the motherland.

On Oct 19, 1952, Princess Elisabeth and the Duke of Edinburgh were visiting Canada and the royal train was to pass through Notch Hill. Grandma and Grandad and the whole family went down to the Notch Hill station to wave as it went by. They were surprised and elated when the train chugged to an unscheduled stop and Princess Elisabeth came out on the platform. With hardly anyone around, Grandma seized the opportunity to talk with the princess. Upon seeing baby John, the princess asked, "May I hold him?"

Grandma beamed with pride as her two-year-old son clung to the smitten future Queen of England.

Grandma grew up on a farm in Hobson, Manitoba. The youngest of her clan, she carried the mantle of family milker until she left home at age twenty-one. Grandma knew how to farm, and Grandad would figure it out. The property came with a couple of Jersey-cross milk cows and, within hours of taking possession, they began the twice-daily routine of milking by hand. In the 1950s, a couple of good dairy cows could support a family by selling just a few gallons of cream a day. They gradually increased to a herd of thirteen and were truly thriving.

They had two babies there – my mom, Joyce, and my uncle John. Their little farm life could be fairly described by any episode of Little House on the Prairie. Grandma milked the cows, fed the chickens and the pigs, prepared all the food and tended the house while carrying around two little kids. And Grandad worked the fields, logged and milled lumber, dug wells, did the heavy lifting, and generally developed the property.

Over the next five years, they bought a few more adjacent pieces of land and developed them as well. They had no debt, they mostly ate delicious, healthy, homegrown food, and they continued to save up money from the sale of their farm produce. In what would be a miraculous feat today, my grandparents started with no formal skills and no money and, within seven years, owned a money-producing, food-producing family farm with over 100 acres of timber, pasture and crops. Owned. Not held with a mortgage and amortized over thirty-five years. They would go on to sell the farm and follow Grandad's heart for the north coast, but they would never need a mortgage. They lived their whole life humbly, within their means and without debt. Their wealth grew year after year. They never pursued financial wealth and, while they owned all the assets in their comfortable lifestyle, that's not the kind of wealth I'm talking about. Even in their 70s, they grew huge gardens and ate fresh and healthy foods. They raised two loving children who, in turn, gave them seven grandchildren and twenty-two great-grandchildren. They played horseshoes, cross-country skied and went trout fishing with their friends and family – all the things you would pay a million dollars for if you didn't have it. But all the money in the world can't buy what took them a lifetime to grow and foster.

Grandad never had a typical career. After the farm, he took on various types of work in a handful of different communities where they lived throughout BC. They were never prestigious jobs and never exceptionally well-paying, but they were always suitable for the need and fulfilled Grandad's interest to try new things.

When I was ten or eleven years old and they were in their sixties, they still had their hands in an eclectic mix of humble labour and I would often help them with small janitorial contracts – sometimes two or three in one day.

I used to love it when Grandad took me for a long walk or bike ride along the highway to pick up beer and pop cans. I remember him teaching me to look for French on the bottle even from a distance – then we knew it was Canadian and paid a deposit. Beer tins were worth double a pop tin, a premium $.10. I got to keep the money from my cans, and earning a few dollars of my own when I was just ten was pretty special.

Grandma and Grandad worked their whole lives and I don't think they regretted it. It paid the bills, connected them with the community, and it kept them productive. I think they understood that humans need to be productive, and in the absence of work, no amount of leisure or entertainment could fill that gap. And that's good news, because it's hard to be productive and not make money. What you produce will vary in value to others, but it will be worth something nonetheless. If we view our own productivity as a necessity similar to food and shelter, then an interesting inversion in logic happens. We will get paid to fulfill our own need for productivity.

Isn't that fabulous? I was minding my own business just trying to produce something good, and boom, cash was shoved in my face. I don't know if this was ever an epiphany for my grandparents, but it was life-changing for me.

After living in Hudson's Hope for about a year, I had a burning drive to finally own my own land and get out of the BC Hydro-owned rental. We had made good progress in our first five years of marriage and we were surrounded by beautiful, inexpensive pieces of land. Yet, for some reason, we hadn't managed to save up any fat wads of cash. Maybe it was the single income, still low on the pay scale. Maybe it was the challenge of supporting a family of four. Maybe it was the wonderful Canada Revenue Agency demanding the lion's share.

Whatever it was, we were disappointed to have only a few thousand dollars to our name. Our plan was to save up money, buy outright, and never have a mortgage. We crunched the numbers and it didn't look good. My annual salary was about $40,000 and my take-home cash after federal and provincial tax, CPP, pension contributions and union dues was closer to $30,000. Even if we could put aside $5,000 a year, it would take us almost twenty years to amass one hundred grand.

Optimistically, my salary would go up and we could earn interest along the way, so maybe we could save the necessary loot in ten years. But we were renting a crappy asbestos-contaminated row-house in town with no private yard at all. There was no way we were going to live in these conditions when we were surrounded by the expansive call of the wild.

What we needed was an old chicken coop on the quiet corner of an industrial lot, but the cheapest thing we could find was a neglected mobile home on five acres for $92,500. Here we were, 1,200 kilometres from the city, in a tiny outpost town, but still trapped by the system. We were either stuck paying rent forever or would sign our life away for a mortgage.

Surrounded by millions of miles of crown land, technically owned by the people of Canada, it seemed crazy that I couldn't just buy five grand worth – just a little chunk, nothing special. Just enough to get started. I could whip up a little house in no time – single-level, stick-framed and metal-clad, just like the duck-poo house from six years ago. And by "just like" I mean simple, square, structurally sound and cheap. We would improve our little piece of Canada with our natural pride of ownership and save money without paying rent. It's what everyone starting off needs. Just enough to stay warm and dry and then allow our human creativity and ambition to build towards something better. But that appeared entirely impossible.

We had a look at the trashy trailer and it was straight out of the A&E reality show, Hoarders. It had been rented for years and the most recent tenants had disappeared, abandoning the property like an abused pet. The yard was overgrown and concealed vast amounts of trash and apparently misplaced housewares. The woodworking shop had an actual mound of trash over six-feet tall and twenty feet in diameter. The local transfer station was about ten minutes away but to dispose of their own garbage was clearly too much of a burden.

The mobile home was outdated and smelled funny. The kitchen had the original linoleum that peeled up in a few locations and captured the dirt. In the centre of the yard was a conspicuously placed sheet of plywood with a tire on it. We dragged it back, revealing the crumbling, wood cribbing of a shallow and collapsing well – about fifteen feet down to the surface of the water. Perfect for small children. The realtor who showed it to us started off apologetic and, with every new discovery, hung his head a little lower in embarrassment.

Rose, understandably, saw five acres of trash, but I could see the future and it was awesome. The mobile home, while fundamentally still being a mobile home, was actually not bad structurally. It had triple-pane sliding windows and six-inch thick insulated walls. It was the polar package. It had a sloped roof with shingles and not the normal curved metal. The shop, which apparently doubled as a garbage dump, was twenty-four by forty with ten-foot ceilings. It was insulated, had a metal roof and a wood stove. It had an attached single-vehicle carport that could easily be closed in and insulated. The well was a death trap but it meant there was water. The remaining four acres of the property were totally untouched forest, full of spruce, pine, birch, and poplar.

We could make it beautiful.

We had never imagined ourselves with a mortgage, so we contemplated that reality for the first time. First, we would need help just to get a down payment, as the bank required a minimum of ten percent down. Then, like slaves, we would need to make monthly payments for the next twenty-five years. I detested the thought of it, but at the same time, we were throwing away three hundred dollars a month in rent for a roof and walls in a location I didn't want to be in. I did the math and discovered that our monthly interest payments on a mortgage would be less than our rental payment. And that was the clincher. It was the better of two evils. We offered $75,000 and, with great consternation, eventually agreed to $79,000 – our biggest financial decision ever. Grandma lent us a few thousand dollars to top up our down payment and, in the end, we made payments of two hundred and twenty-five dollars biweekly.

To this day, I resent that there is no reasonable way for young people without family money or an unusual early financial win to become homeowners without significant debt. No amount of working hard and doing without can make it happen. Although it required a violation of my ideology, given the circumstances, I believe it was the only way forward. For the first time, we owned our own land. Kind of.

The makeover was fantastic. It was an ugly duckling transformation story. Working a full-time job allowed evenings and weekends to work our magic. It took a few days of trips to the transfer station before all the trash was gone. Seeing the empty shop and the litter-free yard was an eye-opener.

We brush-mowed the yard, which quickly turned from the abandoned-lot look to something of a lawn. Incredibly, we found over a thousand square feet of carpet integrated with the grass. It had been there so long the grass had grown through its tight plastic weave and created an interconnected matrix of root balls. Ripping it from the earth was not possible with just Rose and I – we had to wait till we had friends to help.

The pattern of cleaning up a disaster on the surface but discovering another layer of disaster below would repeat. Beneath the heap of garbage in the shop, we realized the floor itself had a pronounced sag. Upon further investigation into the tiny, ventless crawl space, we discovered some of the floor joists were just hanging from the plywood floor (they're supposed to support the floor, not be held up by the floor). As we cleaned up old cupboards in the entranceway – a large, single-room addition onto the original trailer – we discovered water staining on the shelves. Our first heavy rainfall confirmed that, indeed, the roof leaked.

We hired a Hudson's Hope old-timer with an excavator to dig out the old well cribbing and put down a new steel culvert. Digging a hole about twenty-four feet deep, we supported the culvert vertically and cut perforations in the bottom ten feet. The old-timer brought in two or three loads of drain rock and placed them evenly around the base. After adding a geotextile on top of the drain rock, he refilled with the native soil. Eight feet from the top, he showed us how to install a pitless adapter on the culvert wall and we ran a new poly pipe and electrical cable back to the house. The new well filled up and, for about $5,000, we had succeeded in securing water – one of the biggest challenges on a rural property. Like so many instances of fixing old problems, we solved one and uncovered another. Before us, the renters and previous owners relied on hauling water from a community well a few miles down the road. With a cistern of 750 gallons, it would take three trips with a 250-gallon tank on a little trailer to fill it up. This took a few hours every week and so encouraged very modest water use. This is still very common in that region, as drilled wells are very expensive and the water quality can be poor. Whether it was our increased water use, or just the passage of time, we soon found the shower drained verrrryy slowly.

We conferred with the same old-timer. He had a kind face, with the slightest, almost imperceptible smile. Thick suspenders held up well-worn denim jeans, accentuating his appreciation for beer.

With a cigarette hanging out of his mouth, almost lighting up his grey stubble, he said "You know. . . I think I installed that septic field. . . maybe twenty-five years ago."

He lit another cigarette and walked the driveway.

"Yup, there it is," he confirmed, pointing at a shallow depression running from the house to a little corner of the front yard.

We had not noticed it before, but now it seemed obvious.

"That's the septic line," he said, "And right here," he paused, "I think we buried an old car."

"Pardon?" I said. "An old car?"

He chuckled, dropped the butt from his mouth, snuffed it out with the heel of his heavy leather boot, and put the freshly lit one in its place.

"Yeah, or a truck," he stated plainly.

He honoured my confusion with an explanation.

"You put an old car in the hole, stick the pipe through a window, and bury it... It works great..." He continued gently, "For a while... See how the lawn is sunk there?"

I nodded.

"Eventually they cave in," he said. "You need a new septic system."

And so our new well ushered in a new septic field. This time, we used infiltrators – magical, interlocking, perforated plastic tunnels. They could drain ten times the volume of typical PVC drain pipe. We went on to use them in all sorts of DIY drainage projects.

Within months, we made believers of even the most skeptical friends and family. They thought we had just bought a trashy trailer, but we'd actually discovered a hidden gem. Like the conversion from sinner to saint, the property went from a discouraging eyesore to charming and inspiring. The redemption of the property was almost spiritual. Our hard work and dedication had made something beautiful where there was once a garbage heap of shame. It was so uplifting to know we could make such an impact. It was encouraging and inspiring and added a permanent new facet to our character. We could make things good.

With the clutter gone and a full woodshed, the first blanket of snow covered all remaining sins of that old property.

One of the things that drew me north was the amount and variety of wild game. Our property was located on the edge of farmland just twenty minutes from town. There was plenty of wildlife and, apparently, someone had been feeding them on our property. We knew this because, as winter really took hold, a small band of mule deer repeatedly showed up in our backyard. We put out a little bit of oats, and quickly confirmed the reason for their attendance. A couple of young bucks and a few does with fawns – they must have been coming around for quite a while because they didn't mind if we approached them. It's not a good idea to domesticate wild animals but, for a few weeks, we felt privileged to enjoy their close company.

Grandma and Abigail on our first property in Hudson's Hope

Rose and Sarah with our first domestic animal, Junior the goat

We got a fair bit of snow that year, and Sarah was finally big enough for every dad's favourite game. I grabbed Sarah with both fists clenched on the front of her jacket, picked her up, and held her eye-to-eye.

"Are you ready?" I asked.

Without waiting for an answer, I lowered her back down close to the ground and to my right side. Then, using maximum dad-power, I heaved her in a perfect parabolic trajectory as far as I could. She went up like a pop fly – arms and legs flailing.

"AAaaaahhhhhhhh!!!"

Dropping about ten feet away, she completely disappeared in a giant fluffy poof of snow. The muffled yells after impact had more of an alarming tone than the gleeful squeals I was expecting. Rose rushed to her rescue, clambering through three feet of snow.

I gave it a shot with Abigail. Just over a year old, surely I could launch her even further, and she definitely wouldn't be able to complain as loudly as Sarah.

"Waaahoooooo!!!"

Poooofff! Abby created her own little crater and disappeared. I don't remember her exact words, but I presume she appreciated it. I do recall Rose was less than impressed. As a young mom, she was fantastic, but she had missed the memo. Fathers must throw their children into bodies of water or piles of snow whenever possible. I didn't want my kids later on in life to be found looking aimless and acting foolish. Wise old grandparents would take notice and say, "That kid's father never threw them in the snow. What a shame."

Spring brought a vibrant burst of bright green birch leaves and fresh grass. We had chickens in the coop and three young lambs and goat kids grazing our front yard along the road. We fenced off a new garden space and tilled it under. We took three hundred feet of pasture fencing and strung it around the trees in the thick forest on the edge of the backyard. It was the perfect place for a pig pen. Watching our girls chasing three little squealing wiener pigs, it felt like our brand-new homestead was starting off with a bang.

With the house, yard and shop already whipped into shape, we started cleaning up the woods that made up the bulk of our five acres. With Grandma's help, Sarah and Abby picked up sticks in the yard. Rose and I cut down bigger deadfalls and built heaping bonfires. Even the forest, with some of the brush and tangle removed, began to look park-like.

Everything we did on our own land was so new and exciting. We built a fire pit enclosed with large stones and circled with wooden benches and lawn chairs. I always loved camping and, here we were – roasting marshmallows, barbecuing with friends and living every day on our own personal campsite.

As if things couldn't get any better on our little piece of paradise, cute baby Julia Joy joined us in July. Rose had done it again – another beautiful healthy baby, born a month before my twenty-fifth birthday. I still had not put much thought into having babies. They just appeared rather naturally. But now, with three kids four and under, we were officially outnumbered. Two kids was definitely at least twice as much work as one, but I wondered if the third would come with some efficiency. It turns out, not much – in fact, being outnumbered might tip the scales to the side of three kids being more than three times the work of one kid.

"How many kids do you want to have?" asked Rose.

"I don't know," was my honest answer. Who thinks about that kind of thing?

Rose had one sister and I had two siblings, so there was precedent for us to stop with a family of five – double the Canadian average of 1.5 kids per family. We were doing our part, but still well within the normal range. What was unique was having three kids before the age of twenty-five. Certainly no one should stop having kids before they're even twenty-five. Most people are just dabbling with puppy "children" at twenty-five.

"I'm not sure, but I don't think we should stop now. We're too young to stop having kids."

Rose agreed, or at least didn't disagree, and we left it at that. Not a mainstream attitude at the time, but the world needs more healthy babies raised by selfless, loving mothers like Rose and by dads who throw them in the snow.

Julia was a very content baby and seemed to be a reflection of the joyful and peaceful homestead we were creating. If Rose and I had been sowing into our freedom plan for the last six years, this summer we would reap the first bounty. I took a parental leave and, as a fresh family of five, we took full advantage of the summer. We gutted the rotten shop floor, enclosed the attached garage and poured a concrete floor throughout. We ripped off and rebuilt the trailer roof, adding a proper overhang and soffits – erasing a bit of the mobile home's stigma and giving it a distinct house look.

We explored and fished remote little lakes nestled in the Rockies. We bought our first full-size truck, a 1997 four-door Ford F-250, that actually fit our family and could handle some of the farm stuff a bit more graciously.

It was at this time I first came to realize that a government-run, unionized organization often doesn't make a lot of sense. Another trainee and I were both wrapping up our program and were ready to bid full-time jobs. We both wanted to continue working full-time at the dam. The

P&C crew there had trouble retaining workers and, even at full staffing, could not complete all the work without significant amounts of overtime. Logically, they would just offer us positions and we would happily accept and provide years of productive service. But something about headcount and policy meant there were no positions for us to fill. Young technologists typically rotated in and then out of remote rural locations like GM Shrum as soon as they could. We both wanted to stay, and yet we would soon be forced to bid elsewhere. Our frontline manager just shrugged and said there was nothing he could do.

With my own personal objectives in mind, but also believing BC Hydro would benefit, I did a quick analysis and wrote a report for the plant manager. The average annual overtime accrued in our department over the last three years was over 4,000 hours. I had seen previous analyses that the dam workers achieved an average productivity of only sixty percent – meaning a full-time worker spent about 1,000 hours a year doing his actual job, and the rest of the time was safety meetings, training, holidays and fluff. The dam could technically hire four extra full-time techs, each contributing 1,000 hours of productive work a year, just to reduce overtime (which was paid at double time). They would save money by hiring the two of us full time and even if, for some reason, the workload decreased over the years, we still had a 2,000-hour buffer before overstaffing would become a concern.

The plant manager seemed impressed and was quite responsive. He said he would discuss it with the regional manager and was optimistic he could make it happen. Just as I was leaving, feeling pleased with myself, he called me back. I was standing in the open doorway of his corner office on the top floor of the GM Shrum control building.

He paused, scanned the room and through the open doorway behind me, lowered his voice, and said, "I don't really want to emphasize our productivity numbers. Just make me another copy of this report but remove the bit about productivity."

I hesitated, evaluating how my altruistic research had just become politicized.

"Ok, no problem," I replied.

I was learning about "optics". And bureaucracy. Triumphantly, our jobs were posted within just a few months. The P&C manager was surprised and congratulated me for making it happen. He seemed sincere, but I wondered why he wasn't embarrassed that a small effort on my part seemed to highlight his impotence as a manager. Regardless, we had just staked a major claim – one that could lead us to the jackpot.

Professional Hunter

With my job secured and our land in hand, it was time to move our focus further into the woods. It was early fall and our friends, Bev and Jay, came for a visit. Jay had inspired me in multiple significant ways, was full of new ideas and not afraid to try them out. A couple of years earlier, they bought a piece of raw land and moved onto it with seemingly nothing. They whipped up a little shed with a wood stove and moved in. They had incredibly found a way to live an adventure and avoid the conventional indebtedness, all while caring for their six-month-old baby, Sierra.

Jay was a hunting fanatic, and I was fast becoming one. He brought with him an assortment of hunting weapons I'd never used before. We set up in the trees just behind the house and tried out his new .50 calibre muzzleloader – a great big long heavy thing. It was easy to expect the kick of an elephant gun, but as black-powder enthusiasts know, it ignites with a smoky poof and gives your shoulder more of a shove than a kick. Like a woodsman's charcuterie board, we continued by sampling his compound bow and .44 magnum lever-action saddle gun.

I have memories as a four-year-old grouse hunting with Grandad. I specifically remember helping him carry the .410 along a fence line between some pasture and the woods. We were in the Smithers area visiting Uncle Buzz. As with many memories that persist from an early age, this one was seared into me from fear.

Grandad suggested we cut the corner of the pasture, going through the bullpen. There was not a chance I was getting into the bullpen. There was nothing but a herd of cattle grazing in the distance, but somehow the idea seemed terrifying. How could he possibly think it was a good idea to cross the barbed-wire fence, leaving nothing between us and non-existent crazed frothing bull? I don't remember if we got any grouse, but I remember hunting. Even at four, I considered myself a hunter.

Rose, on the other hand, was the farthest thing from a hunter. She was scarred from an early age when her pet duck mysteriously and ironically disappeared the same day her mom served roast duck for the first time. She wasn't particularly interested in guns or hunting. But she had no problem eating meat and still, with the heart of a young lover, was happy to join me on any hunting adventure. I convinced her to take the firearms and hunting courses so that, at minimum, she would legally be able to carry a second firearm.

She was very hesitant to fire my large-calibre hunting rifle, a 7mm Remington Magnum, but the cute little .44 Magnum that Jay brought looked harmless. Rose shot the muzzleloader and wasn't too impressed, but the .44, was just plain fun. A real beast in a handgun, the .44 launched from a rifle and supported with two hands and a shoulder, offers a satisfying pop. She shot it once, turned and gave us a surprised and excited smile, and then kept shooting, unprompted. I was grinning ear-to-ear and couldn't have been prouder. She was empowered with her newfound ability and it seemed like the right time to pop the question.

"Do you want to get your own deer?" I asked nonchalantly.

"Sure, if I can use the .44," was her prompt reply.

So, after struggling to stuff my smile inside the truck, we cruised out to the alfalfa fields. We spotted piles of deer, but no four-point mule deer at close range. After Bev and Jay left, taking the .44 with them, we continued to hunt that fall and she would eventually get her first buck with a borrowed 6mm.

Sarah, Abigail and Julia very proud of Rose and her first buck

It was an exciting time, meeting new people and making a lot of new friends. One family in particular would turn me from an amateur into a hunting professional.

We met the boys while volunteering in a church youth program. Eventually, we got to know their parents and hit it off. They had recently moved to the area from the States after buying a guide outfitting territory. Until just the last couple of years, I had exclusively hunted under the direction of my brother Shawn and Uncle Dave and didn't consider myself a particularly savvy hunter. But Paul was desperate for hunting guides and somehow they convinced me that I could be a pro. I had hunted moose in the summer before, we just snuck around the woods quietly and bumped into them. I'd never called them in the rut or had to identify specific antler configurations. And I had never seriously hunted elk. I mean, we had looked for them, but I was not an elk hunter.

I didn't own any camo or any hunting-specific clothing. We had always relied on old wool pants from the thrift store and some form of flannel jacket. There was nothing wrong with the wool pants but nothing says, "beer-league hunter" like a flannel jacket. Paul and Cheri were unfazed and I soon had my assistant guide license.

"Fake it till you make it," sounds like a sinner's game and I don't really abide by it. That said, you could probably accuse me of "projecting confidence." I invested in a beautiful pair of full-height leather and Gore-Tex Irish Setter Elk Tracker hunting boots. They cost more than my first car. I bought my first camo – a matching pair of Gore-Tex hunting pants and jacket from Cabela's. Just like that, I appeared credible.

I had never hunted elk in the rut, but I knew they were way more finicky than moose. I watched and re-watched a stack of Primos elk hunting videos and practiced my bugling in the living room until I became Will Primos. And that is exactly how I became a professional guide.

These hunters were paying quite a bit of money – some of them blue-collar workers who had saved up for years. Others were rich and travelled the world hunting every imaginable species. Pretty well all of them were prolific hunters. I imagined it'd just take a few minutes for them to see right through my façade and demand their money back. My brother Shawn, who had read a book in high school about a sheep hunter who turned on his guide and tried to kill him, warned me to bring lots of extra ammo.

I was never attacked by a hunter, but the season would end with my own bloodshed.

It was the first morning of the first hunt and Paul and I were guiding two brothers hunting together from the Beaverbrook cabin. The brothers had moose and elk tags but, being early September, it was prime elk rut, so we would target them first. Hunting is a lot of just walking around quietly, and that's how that first morning started. So far, it seemed easy to convince Stephen (not his real name) that I was a professional hunter since I was able to walk around quietly quite well.

Thankfully, we heard an elk bugle down by the river and made our way in that direction. There was no question I knew exactly where the elk was, and I was taking us right to him. I had to start calling and this is where I put it all on the line. I bugled, waited and thought, if this guy doesn't bugle back, Stephen's gonna know I'm a fake for sure. But wouldn't you know it? I got a bugle back. We called back and forth and closed the distance – Stephen was excited and I was impressed with myself. Unfortunately, that's all that happened. We pushed a little too hard and the elk just ghosted us. I assured Stephen that elk hunting was very difficult and this should be expected. He nodded and agreed, and I felt relieved.

It was around then we heard a distant shot. In a fairly inaccessible area, we knew it must have been Paul and the other brother. We headed back to the cabin, emptied our backpacks and headed downriver in the direction of the shot. Not really hunting anymore, we made our way quickly, chatting as we walked. We were making no attempt to be quiet – just stomping through the woods. I stepped on a good-sized branch and snapped it with a crack, and out of nowhere came a raspy roar that turned into the rising shriek of an angry bull.

Stephen and I froze, wide-eyed.

He was extremely close and sounded mad but the alders were still in full leaf and we couldn't see far in the undergrowth. I didn't have a bugle or regular squeeze-style call. All I had was a fantastically difficult-to-master little diaphragm call. It was just a small reed that you hold to the top of your mouth with your tongue. It's versatile, but requires skill and practice and feels like a one-man french-kissing competition. We had shed our camo gear and were just wearing T-shirts. Why in the world was this elk all fired up in the middle of the day? Elk hunting is typically done in the wee hours of the morning at this time of year – 5:30 AM.

I made some funny sounds with the call, making sure not to look at Stephen. It didn't matter. That bull was mad and he probably would've charged at us even if I had mooed or clucked. He came through the brush and out on the trail in front of us, flanked by cows. He was clearly a herd bull. I was smart enough to have my binoculars and I quickly ID'd him as a legal six-point. This was crazy. My first day of guiding. My first morning of guiding. And we were going to get an elk. Stephen lined up. He was broadside about eighty yards away. It was an impossible shot to miss.

And he missed.

At the crack of the gun, at least a half-dozen elk tore out of there. We looked and looked, and there was zero sign that he'd been hit. Stephen admitted to rushing the shot. How quickly the tides change. That morning, I was the anxious young guide hoping to prove myself and now, just hours later, I had used a little-known midday marching technique and some very unique cow calling to deliver the trophy, only to have the hunter prove himself a novice. He was still kicking himself when Paul and his brother showed up.

Our sad tale quickly became irrelevant as Paul retold their story. It started with three bulls bugling and fighting in the river and ended with the biggest bull going down. We got to work packing out all the meat, antlers, and cape. Overall, it was a fantastic day of hunting success.

With the first-day jitters out of the way for both of us, we could settle in and enjoy the rest of the hunt. We put on ten to fifteen kilometres every day, slowly covering ground. We bumped into a few moose and elk, but it wasn't until the sixth day that we saw another big one. We hiked down to a new spot along the river in the dark and, at dawn, heard a bugle from the other side. A couple of cow calls, and he came charging out to the middle of the river. It was easy to count – a beautiful 6x6 standing in the river at one hundred yards. Steam was rising from the water and the sun was just igniting the treetops with its first golden rays. It was a picture of majesty.

Stephen had no idea when he shot that it would be the first successful elk hunt I was part of. He was pumped and we both felt like heroes. We worked for a couple hours skinning and quartering and then a couple more packing the first load back to the cabin. Paul seemed equal parts impressed and relieved. If I had been "projecting confidence" in myself, I think he had been "projecting faith" in me.

The next hunt, it would just be me and Eldon at the cabin. We arrived early and had a couple of hours of daylight left. We decided to sit at a little tree stand for dusk and look for moose. We sat there for about forty-five minutes and nothing happened. Without any explanation, I turned away and let out a long moaning cow moose call. I wasn't sure if Eldon had ever heard a moose call and, when I turned back to look at him, I could tell he hadn't. He just stared at me, and it seemed like he was deciding whether to pretend it didn't happen or slowly try to back away. If you've never heard a moose call, it's a nasal-sounding moan with a bit of vibrato. I'm quite pleased by the sound but, objectively, it's not pleasant or attractive. I could tell he was hoping that I was joking and not actually a psycho. And that's when we heard a bull moose grunt back.

A low, guttural sound, barely audible at a distance, it conjures an image of a constipated wombat more than that of a half-ton antlered beast. Still, when any animal speaking a foreign language seeks you out, it kicks off an adrenaline rush. Eldon was wide-eyed — first from my own awkward moaning sounds, then the bull's response, and now, a second bull. The two of them emerged from the low brush in the middle of the clearing we were sitting over.

I'm consistently amazed at how large game can hide in plain sight. The bulls made their way in our direction, crisscrossing paths. I could tell one was too small but the other one was very close. It took a minute to confirm it had three brow tines on one of the antlers, making it a legal mature bull. Eldon was beside himself with excitement. He'd never seen a moose, let alone two grunting bulls coming his way. I let him know that the closest one was the smallest legal bull I'd ever seen and he immediately shot it.

The thing about a bull moose – even a small one – is it's really big. I wasn't sure what Eldon's response would be but, as we approached the dead animal, he was in obvious awe of its sheer size. Maybe five years old, it was probably 800 pounds live weight.

We set to work skinning and quartering and by midnight we were loading the meat on my old '95 Yamaha Kodiak, one of the bulletproof early model ATVs. The quad could basically haul anything you could strap onto it. With two quarters on the back and two in the front, plus the head and antlers, it handled the 500 pounds with ease. I suggested I could take the meat out to the truck and come back for Eldon and our gear.

Eldon gave a slow, "Okaaaayyyyy, or I could just come with you."

We had doubled in together and I already felt low-budget. But snuggling together with a moose was next-level unprofessional. Of course, what I thought didn't matter. Eldon was the client and clearly didn't want to be left alone. So, with an entire moose and our gear, I climbed on and slid as tight as I could to the handlebars. Eldon initially tried to sit on the meat on the rear rack but soon succumbed to gravity. He slid down nice and tight, making the perfect moose-Eldon-Jeff-moose sandwich.

With over 1,000 pounds on the quad, I could've been concerned that it might break on the bumpy trail but I was much more concerned that someone would see us. Thankfully, it was two in the morning in a remote corner of Canada and no one was there to collect photo evidence.

We returned to the main lodge victorious only twelve hours after we left. A bit of a sleep and a big breakfast later, we regrouped and headed back, looking for elk. Peak rut was over, but on day three we managed to call in a nice five-point within three yards. He had come from more than a kilometre away, and we had watched him walk up the river and then up a steep bank towards us. I had been cow-calling but stopped when we first saw him a couple of hundred yards away. Even without any additional signals, he knew exactly where the sound had come from and eventually stood right over us. We were sitting under a spruce tree, leaned against the trunk, and the bull approached, stopping just at the tips of the branches.

His chest was heaving and saliva and mucus were dripping in a stream from his mouth. He was nervous and the muscles along his flank quivered and twitched. In a seriously compromised position, it was impossible not to imagine the fatal carnage he could inflict with his sharp hooves and massive rack. If elk were carnivorous, we would have been goners. If we just startled him, he could stomp us with a flurry of hooves or gore us to death out of pure reflex.

Thankfully, Eldon knew what to do without any direction and we both sat silent, motionless, and breathless for an everlasting thirty seconds.

The bull's eyes grew more wild as he realized the trap. In an instant, he exploded into motion – dropping to a powerful crouch and spinning at the same time. He launched himself back down the hill and, in a second, was gone. Eldon never did get an elk. I'm not sure how he felt, but I certainly value that close encounter as much as a successful kill.

On the way back to camp that evening, we hiked past the place where he had shot the moose just days before. We had left a couple of hundred pounds of gut pile, bones, and hide, and they were totally gone.

"I'm sure it was right here," I said to Eldon and he nodded in agreement.

As we inspected closer, it became obvious that the ground had been raked – not just lightly raked, but raked right down into the dirt. It was almost totally dark, barely enough light to follow the trail. I switched on my headlamp and instantly recognized the situation. I was standing on a grizzly bear cache. Not able to eat the whole carcass, he buried it by raking up all the grass and debris and dirt until it was entirely covered. I immediately started searching the shadows for any sign of the bear, knowing he would not be far. Satisfied that there was no imminent charge from the bushes, we raised our guns and made a tactical retreat.

Back at the cabin, we met with destruction. We circled carefully, guns drawn like the SWAT team at a drug bust. We weren't totally surprised. One of our first scouting trips to the cabin in the summer revealed freshly chewed and torn rafters, together with mud and claw marks the height of the wall. If your cabin resembles an eight-foot-tall scratching post of shredded lumber, you might be in grizz country.

The casualty this time around was an eighteen-pack of Pepsi we'd left on the porch – an unnecessary indulgence to help celebrate the world's smallest legal mature bull moose. Once we were satisfied that the bear was no longer in the immediate area, we allowed ourselves to imagine exactly how it all went down.

Bears are curious and, like babies, they discover things by putting them in their mouths. So it's no surprise that, even though it had no food smell, the grizz would tear up the box and bite a can. I wish I could've seen his response as it exploded in his face and sprayed in his eyes. He would've jumped back and stood on his feet and swirled around, looking for the sudden source of irritation. Then, once over the surprise, he'd lick his lips and fall in love. He returned to the porch and devoured seventeen more cans of Pepsi. He actually ate most of the cans.

We picked up all the shredded pieces and estimated no less than half the tins were totally gone. The wall of the cabin, the door and the whole porch were stained sticky brown. It was a one-bear frat party, popping Pepsi like a one-year-old eating birthday cake.

After two successful hunts and the better part of a month spent in the woods, I had completely embraced my new life. I was in great shape, easily hiking six to ten miles a day, spending one hundred percent of my days outdoors and swapping stories with other hunters. It was the immersive wilderness adventure I always wanted.

I could finally identify all the local species of trees and some of their various uses. I could usually find edible berries, which were always a welcome boost when wandering in the woods. Birch bark and spruce sap were like rocket fuel when starting a fire. Alder, due in part to its wide availability but mostly because of its combination of strength and flexibility, was the go-to wood for general camp craft. A great big spruce growing on its own will grow such broad thick branches, they could keep you dry in the most intense downpour.

Guiding had finally allowed me to peek underneath the kimono. It was early days, but I was taking my relationship with the backcountry to the next level.

A Bloody Joke

My last hunt of the year was the quickest and most memorable. It was a moose-only hunt and Paul and I were again hunting in the same area, each with our own hunter. The first morning, I was sitting in the same tree stand that Eldon had shot his moose from ten days earlier. It was still dark, but I could hear something rustling at the far edge of the meadow. In the earliest dim of dawn, I could see a big set of antlers shaking around in the brush. I was glued to the binoculars for half an hour as we approached shooting light, and the hunter's anticipation was growing. With his rifle trained in that direction, we both waited as the light slowly showed us the outline of a great bull moose.

The shot rang out the first minute of legal light and the bull instantly disappeared from sight. We watched for a few minutes to make sure we hadn't missed him sneaking away and then climbed down and marched to where he had stood. The shot was about two hundred yards and, with a good rest, should have been no problem. But we couldn't find him. We were certain we were standing on the small rise the bull had stood on, but still we couldn't see any sign. Behind us was a beaver channel and a small pond. There was certainly no moose in the pond, but there was a little clump of branches sticking out that looked a little suspect. We walked around to have a better look and – wow – ninety-nine percent of a 1,000-pound beast was totally submerged and hidden from view with just a couple of inches of antler tines sticking out.

Paul and his hunter were still heading to their spot in the morning when they heard the shot and turned around to find us. They showed up at the perfect time and the four of us and a quad winch got him out. The hunters were pumped, and Paul appeared impressed and less surprised at another early success. We had the whole day and all four of us to do the work. It would be a breeze. After taking some photos and a little jovial banter, I set to work skinning. As soon as I brought out my Buck knife, my hunter raised one eyebrow and commented on its grand size.

"Do you really need such a large knife?" he asked.

These were deer hunters and had never hunted moose, so I could've just told him, "Yes, you need a giant knife with a seven-and-a-half-inch blade to gut a moose."

But I told him the truth, and that was, this knife was my backup protection in case I found myself being mauled by a bear. Not allowed to carry handguns in Canada, it's the only option available if a bear gets too close. I certainly didn't like my odds of fighting a bear with a three-and-a-half-inch skinning blade.

I started with the tough hide on the lower leg. I was applying a lot of pressure to try and break through the hide and realized my left hand was supporting my weight further up the leg and was in danger if I slipped. I wisely re-positioned and continued to put all my weight behind the knife. Without the support of my left hand, I was a little unstable, and so, when the knife inevitably slipped, I lost my balance. Instinctively, I thrust out my left arm to break my fall. All in the same motion, the knife in my right hand T-boned my left forearm and sunk straight in. Reflexively, I pulled the knife out, dropped it and clasped the wound with my hand.

There was about an inch and a half of blood on the end of the knife. I looked up to meet the stare of Paul and the two hunters.

It turns out Paul is very queasy around human blood and his wild eyes betrayed it. His manner turned instantly tense and he demanded to know how bad it was. I knew it was deep, but I didn't know if it was "bad." I lifted my hand as if just taking a peek at my poker cards. What I saw did not look like other cuts. It is best described as a gaping meat mouth – a crevice of bloody meat. If it had a few teeth, it might've looked like the gaping mouth of a boxer after being beaten in the face.

"I don't think I'll be able to help with the moose," I said, closing the wound back up.

Paul was totally unsatisfied, and his body language was getting frantic. He was looking around as if he'd lost something. It was funny to watch this big, tough hunting guide – the boss – dissolving into a pool of anxiety right in front of us. I remembered the clients, who were calmly waiting to see how this played out, and figured I'd better calm Paul down. I repeatedly assured him I was fine.

I had clearly not struck an artery, or blood would have been spurting everywhere. And it didn't seem like I'd severed any tendons, as my fingers and hand still seemed to move OK. So it was just a big muscle cut – I would just hold pressure and they could continue getting the moose skinned, and, when we took it back to the lodge, they could take me to the hospital. I would definitely need stitches.

Paul chilled out a bit and they all went to work. As soon as their attention turned back to the moose and I could collect my own thoughts, I couldn't get over how emotionally affected Paul was. I was grinning to myself and trying not to laugh out loud. Paul was a true alpha male and to see him exhibit signs of weakness from a little flesh wound was hilarious. I wanted to poke the bear a bit, so after a couple minutes, and with everyone settled down, I told them a story.

Mustering my most serious tone, I said, "Paul," and he looked up. "I'm totally fine," I reassured him again, "But there actually is a small chance I could lose consciousness."

Both hunters stopped what they were doing, and all three gave me their full attention.

"I could lose consciousness. Probably not, but it is possible," I stated blandly.

The hunters seemed surprised but just continued to stare and didn't say a word. Paul, on the other hand, instantly got that crazy look again.

"What?? Why? What do you mean?" he stammered.

Knowing I was onto something, I suppressed my inner glee at his response and continued.

"If, for any reason, I lose consciousness, I'm gonna need you to do something. In the bottom of my backpack is a first aid kit, and inside, there's a giant needle."

Paul was turning pale and his pupils looked like they were going to vibrate right out of his eyeballs. I could tell his mind was going a mile a minute, but the only word he could spit out was an incredulous, "What??"

I repeated it a little slower this time.

"There's a giant needle, and you'll need to stab it straight into my heart."

I looked at the hunters, who were speechless with disbelief, and then slowly panned over to Paul, who was clearly on the edge of a meltdown. I remember being so proud of myself, keeping a straight face in the midst of such a wildly successful prank.

Having lit the fuse, I took a seat on a log, pretended nothing had happened, and prepared to watch the fireworks. I would let Paul freak out and embarrass himself in front of his clients, and then slowly let him in on my little joke. He'd be mad, but one day he'd get over it, and sitting around with a bunch of guys, would humbly acknowledge that I'd one-upped him.

But that's not what happened.

I sat down on the log and everything got a little blurry. The bright sky of early morning got noticeably darker and little flickering and flashing lights appeared all around me. I could vaguely hear Paul yelling at me but it got real quiet and it felt like my ears suddenly plugged. I could just hear the sounds from inside my head and then it went totally dark.

Someone was shaking my head and speaking loudly right into my face.

I felt someone strike my cheek.

"Jeff, Jeff!"

I could feel dirt in my teeth and raised my hand to my face and felt mud.

"Jeff!" the guy said.

"Who was this?" I thought to myself.

I was staring directly into someone's face, but I had no idea who it was.

"Jeff!"

They clearly knew who I was, but... and then it clicked. The guy was my hunter. I was lying in the dirt and he was kneeling over me. The other hunter appeared and pulled me up to a sitting position.

"What's going on?" I asked them.

They didn't say anything. They just stared. I was genuinely confused. It seemed a bit confrontational to hit me and yell in my face, so I asked again, "What's going on?"

They both looked skeptical but slowly explained.

"You fell on your face. You fell backwards off the log, but kind of rotated on the way down and just landed on your face."

Finally, I remembered. Through the brain fog, I quietly surveyed my surroundings until my eyes landed on Paul. He was vibrating like the Energizer Bunny in a straitjacket and he was rooting around in the bottom of my backpack. It all came rushing back and I suddenly realized what happened. Never in my wildest dreams could I have intentionally pulled off such a great stunt.

I burst out laughing and, suddenly relieved, the hunters joined me. Paul was hopping mad and bewildered as he rebuked me.

"When you first fell, I just thought you were joking," he explained. "But you didn't just fall back – you twisted up super awkward and smashed your face straight into the ground. Who would fall right on their face just to make a joke?"

I could not respond with words and just laughed harder. We were all laughing now – even Paul.

He leaned in and, with all seriousness, said, "You're lucky there's not a giant needle in your backpack."

Aunt Joy always drew a very clear distinction between a mountain man and a bushman. A mountain man packed everything on his back and ate noodles cooked on a WhisperLite camp stove. He slept in a nylon tent and left nothing but footprints. A bushman often drove a big truck. If a tree got in his way, he cut it down. He cooked up lasagna his wife had prepared and packed for him. He slept in a wall tent with a wood stove and he left behind mud-terrain tire tracks in addition to his footprints.

Aunt Joy was all mountain man(woman) and, as long as I can remember, she taught my cousins and me the skills of surviving and enjoying the wilderness. In preparation for a multi-day hiking expedition, we would carefully prepare and weigh all our food. We would not bring any extra. We brought two pairs of thin liner socks and two pairs of heavier wool socks. We wore fancy polypropylene long johns and undershirts – cotton was not allowed.

She would say things like, "Hurry up, you little potlicker! If you took this long to make camp in a blizzard we'd all die!"

Each summer when I was seven, eight, and nine, we hiked in Manning Park to the Three Brothers and Nicomen Lake. When I was eleven and twelve, we hiked the 80-km-long, wet and rugged West Coast Trail. The first year was cut short by a medical emergency but the second year was a successful hike through. At fourteen, we hiked from Canada's hotspot in the Fraser Canyon up the Stein River Valley. On all these trips, the stunning and natural beauty of mountain peaks, rolling alpine, crashing coastline and glacier-fed waters was revelatory. We learned to camp with low impact and respect wild places. We learned the importance of working fast and heeding the weather to avoid disaster.

Aunt Joy taught us well.

"Sometimes you just have to shut up, eat your gorp, and keep hiking!"

We all loved Auntie Joy, have terrific memories of those trips and adopted the mountain man way into our character.

Uncle Dave, Aunt Joy's brother, conversely, was the quintessential bushman. Starting at age nine, I would spend a couple of weeks with my big brother, Shawn, and Uncle Dave hunting moose in the north. Uncle Dave had a '77 Ford F-250 Highboy 4x4 he called Old Blue. Decades of tough love and, sometimes, straight-up abuse had left it with war wounds on every body panel and a little rust on the fenders. At first glance, it looked like a fixer-upper. But look closer and you'd see the heavy, custom, steel front bumper with a trusty old Warn winch. Its stock 33 x 7½-inch, heavy-lugged, pizza-cutter military tires were perfect for cutting through marsh grass and muskeg to reach firm ground.

Stowed under the old-school, fibreglass sleeper were all the necessary tools for tackling the woods – chainsaw, high-lift jack, wrenches and sockets, tow ropes, tire chains, come-along, tire patch kit and spare spark plug. Under the hood, the gas-guzzling, stock big block had been removed, replaced with the bulletproof 300 cu. in. straight-six. The straight-six was one of the most reliable engines ever built and, with its high-torque at low-RPM, it was perfect for chewing its way through the boreal forest.

As fun as the hunt was, the epic journey getting there was already worth the price of admission. We would start on the gravel roads and quickly turn off onto old oil and gas exploration trails which were cut through the woods thirty years earlier for winter drilling. They were never intended to be reused. Providing mostly inspiration, we would loosely follow these old cut lines but often navigate our own way around muskeg, washouts, and creek crossings. Chaining up all four was required almost immediately and most obstacles were overcome with a steady throttle and flinging mud. There were always a few gnarly scenarios that required a campaign of winching and road-building. We would cut dead windfall to pack in muddy ruts.

The smell was a unique and memorable combination of sweet pine sap and mud burning off a hot tailpipe. Uncle Dave never wasted much time overthinking things. He'd have a look and make a decision.

"Well, Old Blue's not gonna winch itself over there."

I would often run ahead to clear the way. Tackling the tangle of brush, pine needles would find their way down the back of my sweaty T-shirt. Muddy water would slowly wick its way up my jeans and the churning tires would inevitably coat my face with mud as the truck chewed through the marshy ground.

It was everything a boy could hope for. Sometime, mid-afternoon, after extracting ourselves from a particularly steep and slick creek crossing, we would sit on the moss in the shade of a spruce and have a peanut butter and honey sandwich.

We were conquering the wilderness.

Camp would often include a wall tent and a 2 x 8-foot sheet of plywood nailed to the poplars for a kitchen table, a two-burner Coleman stove burning white gas, and a cast-iron pan to fry heaps of bacon and eggs for breakfast. For ten consecutive years, Uncle Dave, Shawn and I – and sometimes others – would hunt moose in August. We usually got a moose and it was a fantastic adventure every time.

I became an untethered half-breed – half mountain man, half bushman. Sometimes one and sometimes the other. Through inheritance and experience, I developed two overlapping and complementary outdoor skill sets. Kind of like the ecclesiastical expression, there is a season to prance around the alpine in polypro and there is a season to cut through it with a chainsaw.

If I was to create the life I wanted, I would need to adapt these skills and add some more.

Bear, Elk and Bison

Everyone was asleep except for me. It was one of the first nights on our new land in Hudson's Hope. It was totally dark out and totally quiet – and I mean that in the context of growing up and living in the suburbs my whole life. There were no streetlights. There was no traffic sound. It was truly a very, very alone feeling.

I remember looking out the big window in our bedroom. It was inky blackness. In my apparent state of deafness and blindness, I suddenly felt a tinge of fear. You know, the kind of irrational fear you have as a child lying awake at night, maybe after watching some scary movie. My irrational, yet natural, response was to listen even harder in the quiet for the subtle but irrefutable sounds of a mass murderer sneaking around outside the house. It was the only time I remember that sensation as an adult. For the first time, I felt truly alone and solely responsible for protecting my family and myself.

There's something consoling in the constantly droning sounds of society playing 24/7 in our lives.

I was twenty-four years old at the time and I remember being embarrassed at my immature and fearful thoughts. In hindsight, I think it was probably reasonable, given my push to advance to the next level of self-sufficiency. The fears were just a brief flashback to childhood, yet they offered me the opportunity to acknowledge and accept the increased fatherly responsibility of removing my family from the suburban comfort zone.

I stared at the window and dared anything terrible to happen. I waited. I listened. And sure enough – nothing. I lay there totally defenseless. No ready weapon. Something galvanized deep inside me. Fear is the enemy. Fear should be faced and rejected.

The whole experience probably lasted less than a minute, yet it had a lasting impact on my psyche. It internally signified a giant step away from city kid and toward mountain man. My family needed me to be courageous, and I was honoured to be their fearless leader.

I didn't realize how soon I would be called upon.

It was mid-fall but we already had a couple inches of snow. We were still getting used to our new routine as chicken ranchers and had recently suffered our first chicken massacre. Dead and dying chickens with blood everywhere – sure sign of a murderous weasel. We learned the hard way why people close their chickens in the coop at night and added this to the nightly chores.

The task of tucking the chickens in for the night fell to Rose.

One evening, Rose suddenly stated, "Oh shoot, I forgot to close up the chicken coop."

"You should do that right away," I suggested.

She looked at me, glanced quickly out the window to confirm that it was now totally dark, looked back at me, and said, "No, I think you should do it."

I knew exactly what this was about. We'd had a bear come through a few times in the last few weeks and she was scared of the bear. He knocked over our trash can one time – which was my fault for leaving it outside – and came through in broad daylight another time, doing a slow walk around the chicken run.

Good advice in bear country is: if they come through once, scare them off; if they come through again, shoot them. The truth is, regardless of who's at fault, a bear that's not scared around humans is a dangerous bear. If I saw him again, I had determined to make good on that advice. But I wasn't about to let Rose get pushed around just by the thought of bumping into a bear.

"Rose, there's no bear out there," I assured her.

She repeated unequivocally, "I'm not going out there."

"Wow," I thought, "She's being stubborn. I can't let that stand."

"Rose, I'll help you – come with me."

I gestured for her to follow me to the door.

"No," she replied and stood statue still.

"Rose," I pleaded, "There is no bear. Come I'll show you."

I took her hand.

"Even if there was a bear, you just tell it to go away." I explained.

As I swung open the front door, I said jestingly, "Hey bear, get out of here."

And there he was, staring at us from four feet away at the bottom of the step. He spun and tore off in a fury, kicking up snow and huffing. He was so close we could hear the jostling of his muscles and joints as he hurled himself into the darkness.

Rose jumped back, exclaiming, "SEE!!"

She glared at me with righteous indignation. She unlocked a new fear but also glowed as if the Lord had just anointed her as chief prophetess. I wasn't totally wrong. The bear did run away when I yelled at him. But I had clearly been outdone. I had beaten the black bear but lost to mama bear.

"Do you want me to close up the chickens?" I asked, and without waiting for a response or making eye contact, I trudged out to the chicken coop.

After my first fall of guiding, I was highly motivated to finally get my own elk.

"You must've hunted elk so much, you get tired of it," the hunters would say.

I would reply with something like, "Oh, it never gets old for me."

Which was true and sounded better than, "I've actually never shot my own elk."

We lived in ranch country, surrounded by the perfect mix of timber, pasture, oats and alfalfa fields. The place was literally crawling with elk, moose, whitetail and mule deer, grizzly bear, black bear and wolf. I had permission to hunt about ten sections of land just ten minutes down the road and spent evenings and weekends there all October.

There was a resident herd of elk and, after the rut, the bulls gathered in a bachelor group. Day after day, I hunted the same seventeen bulls but they were crafty and never held to a predictable routine. I would set up on one side of the field and they would come out the other. The following day, I would sit across the field and they'd come out a mile down.

It was the last few days of the season and Jay was back up. He had a bison draw further north and Uncle Dave and I were going with him. But first, he offered to give me a hand with these elk.

It was frosty and clear as dawn lightened the sky and Jay and I were tucked against the bush, hoping to get a glimpse of the bulls. Sure enough, they came out about five hundred meters away.

We lowered into a crouch and crept towards them.

The big bulls had paired up and were sparring, which made counting points almost impossible. At least half of them were either five or six-pointers, but to confirm and then keep track of an individual bull was proving impossible. They worked their way across the small field. About three hundred yards away, they started hopping the fence and disappearing into the woods one by one. They were slipping away, and so I put all my hopes in the last big bull that would slip from sight.

With just a second to spare, I counted six and Jay and I both shot at the same time. The bull crumpled. We ran up, excited but also nervous. It was always a possibility I had miscounted or made a mistake and I couldn't imagine the thought of shooting an illegal animal. Would I turn myself in and lose the animal and maybe my license? Would I just keep it and eat it and always feel shame?

The trepidation turned to celebration. Jay and I had shot our first six-by-six bull elk together. Giant and beautiful, we were about to find out just how good elk meat was.

As if successfully harvesting my first big bull elk wasn't good enough, we were about to embark on an epic journey into the Serengeti of the north, looking for bison. We had all been putting in for the bison draw for years and this was the first time anyone I knew had won the tag. Thankfully, Jay enlisted help from Uncle Dave and me and we embarked up the Alaska Highway to the eastern slope of the Rocky Mountains.

The Muskwa-Kechika Management Area is the epitome of everything beautiful and wild. It has deer and elk and moose, but also sheep and goat and caribou, and of course, bison. The boreal forest on the east makes a sudden transition into rolling, alpine-topped foothills. Poplar meadows intersperse with thick mature spruce forest in the river valleys and up the steep slopes. Like the Peace Country, there's a healthy amount of alder and willow, but the underbrush is mostly low and grassy, with spring-fed creeks trickling out of the hillsides, combining with glacier-flood-fed rivers. The whole area is like a garden for North America's largest beasts.

We'd been there before and watched a grizzly bear tear after a moose for miles through meadows and timber while, at the same time, glassing sheep and elk on the opposing slope. We've seen herds of elk on the very top of the mountains and watched our first wolverine hobble around in the valley below. Anything is possible out there and, with the confluence of rolling alpine, rocky peaks and broad river valleys, there is no more majestic setting.

This time would be different. An early winter hunt, the vibrant colours had given way to a monochromatic white and gray. We loaded up my two quads and Uncle Dave's three-wheeler with a wall tent, woodstove, and ten days worth of gear. We were prepared for cold weather and easily handled the chilly -15°C on the ride in. We arrived two days early and set up camp, leaving one full day for scouting.

There was four inches of snow on the ground and bison tracks everywhere, so we just followed them. A few miles down, it was clear we were getting close to the main herd. The tracks turned into a pounded path littered with warm bison patties. We snuck in the last few hundred yards and edged through the brush into a clearing packed with bison. There were probably one hundred or so – cows, calves, and bulls. They looked pretty content, so we backed out and headed for camp.

In November up there it's dark around four in the afternoon and not light again until nine, so we had plenty of time to cut firewood, stoke the stove, and mull over our strategy. One of our most complicated, it went something like this: "We'll get up and hike to the clearing in the dark. When it gets light, we'll shoot one."

One of the rare occasions when a hunt goes to plan, dawn found us lying prone, side-by-side, crosshairs on a nice bull. It was Jay's tag, and so he fired first with a borrowed .300. It looked like a good hit but Dave and I were here to make sure it did not get away.

Bison are known to eat lead pills and continue charging miles into thick timber. Dave shot with his .300 Win Mag and I shot with my 7mm Rem Mag. I couldn't imagine we missed – it was only one hundred yards – but the whole herd was moving quickly into the trees and our target was trotting along with them.

Uncle Dave took aim again and landed a perfectly placed shot in the neck, dropping him instantly. He was dead by the time we cautiously made our way up to him. The first two shots both hit him square in the lungs, and my shot – a little bit high and further back – hit him in the liver. All of them were kill shots but I'm not confident they would've stopped him before he got out of sight.

Either way, we now had the eye-opening experience of cutting into our first bison. We skinned carefully to preserve the entire hide and then quartered and broke down the quarters. In total, we had ten separate pieces, totalling in the five hundred to six hundred pound range. It's easy to see how the bison sustained First Nations and Métis on the prairies through the harshest winter conditions. The hide was fifty to sixty square feet and the shoulder mane was nearly a foot thick of woolly insulation. The meat from one animal could feed a village for a week.

These were the adventures I had dreamed of and I wanted to include Rose and the girls in them as much as I could. Early the next season, I surprised Rose with a babysitter and whisked her away on the quad. Our place backed directly onto Crown land and immediately behind the property was a little cutline perfect for walking or quadding. We rode about a mile and a half to a little tree stand I'd set up. We sat twenty feet up the tree, romantically squished on a single seat. She had a rifle and would shoot any little whitetail buck we saw, and I brought the bow. It was a beautiful, mild evening – clear skies punctuated only by the sounds of songbirds. Thinking there was a much better chance of just kissing in a tree than actually seeing anything, I was surprised when a black bear rolled in. I seized the opportunity and poked him with the bow. It was our first proper date in years, just Rose and I in the sunset, elbows deep in a carcass.

As I slowly internalized the fact that I could hunt out my back door, it occurred to me that I could probably sneak in a quick hunt before work in the morning. In early September it's light at six a.m. and I didn't have to leave for work until twenty to eight. I set out in the dark, quadded a mile down the trail and then kept walking.

Within minutes, I heard a bugle and worked toward it. Using the Hoochie Mama cow call, I made a few small chirps and then stood totally still and listened. This bull was ready for action and screamed back immediately. Doing my best to estimate how far away he was and exactly which direction, he negated my calculations by busting through the brush at twenty yards.

I shot him right behind the shoulder and he went down within seconds.

I ran up to him, astonished at how a quick walk in the woods before work turned into my first solo six-point. A beautiful giant beast, he would feed our growing family for a year.

I glanced at my watch. It was 7:15 a.m. I was somewhere in the woods a mile from my quad, which was a mile from my house, which was fifteen minutes from work. I hadn't brought any ribbon, and I had no phone or GPS. How would I ever find him again?

I didn't really know where I was, but I started in the general direction I had to go, lunging with my left foot and dragging my right foot. I occasionally looked behind to see if I was leaving a recognizable trail. When I got back to the quad trail, I dragged my toe and made a good gouge in the dirt. I ran to the quad, tore off home, hopped in the truck, and got to work – maybe a couple minutes late.

Noting my sweaty, disheveled appearance, my boss greeted me with a look of concern, "Are you all right?"

"Yes, I'm great. I shot an elk! It's still out there."

A hunter himself, he replied, "Well, what are you doing here? I'll see you tomorrow."

By nine that morning, Rose and I were on the quad with two-year-old Abby. She was riveted as we found the trail and, eventually, the elk. Working together to get it out and receiving a hero's welcome when we brought it home to Grandma and the other girls felt like a major achievement. We'd reached the next level of self-sufficiency.

Jeff and Abigail with a before-work six point bull elk

Dream Property

We had totally transformed a literal trash heap into a fine home and food producing property over the last two-and-a-half years. With loads of hours invested on evenings and weekends and about $20,000 spent on all the major upgrades, we had turned a disaster into functional beauty. Combined with a rapidly growing local housing market, our place was now worth over $200,000.

We had our eyes set on a picture-perfect secluded half-section with ninety acres of timothy pasture, with Crown land on two sides and the other two sides bordering unoccupied farm and timberland. It was bursting at the seams with deer and elk and moose and had an unobstructed view of Butler Ridge. There was nothing on it and there were no visible developments on any adjacent properties. The nearest neighbour was a mile away. It did not have proper road access but it did have a right-of-way – we would just have to clear it ourselves.

We picked out a perfect spot for the house and determined it was eight hundred meters from the gravel road and nearest powerline. This was undoubtedly our dream property. It was perfect. We would build our own tiny little off-grid cabin, be debt-free with a huge garden, and we would hunt all our meat and live happily ever after.

So, with Rose pregnant with our fourth, we bought the raw land for $90,000 and put our place up for sale. Our plan went off the rails almost immediately. It was very subtle, and we didn't recognize it at first. But a few seemingly logical choices would eventually have massive consequences.

To buy the land prior to selling our house, we had to get a mortgage through Farm Credit Canada (FCC). A bank mortgage was not an option as banks hate you and don't want you to own land. Okay, that's a little harsh, but really, banks are totally uninterested, unless you have oodles of wealth or commit to a restrictive construction mortgage, they will not lend money for raw land. This is, of course, a massive problem for anyone who does not want to buy a cookie-cutter house at retail price.

We've talked to many people over the years, leading me to believe there are probably millions of people across Canada who would love to buy a piece of land and build their own house. With the alternative being generational levels of debt and, with the example of my granddad in the 1950s, it seems like a reasonable idea. I'm no economist but clearly the withered hand of government has regulated the free market into a pretzel, forcing would-be homeowners/builders out of the land market.

Because the heavily regulated banking sector doesn't want to lend, the federal government legislated a bureaucracy, Farm Credit Canada, to meet the needs of farmers. Classic governmental move – distort the free market until it doesn't work and then "fix it" with more government bureaucracy.

Through the winter of 2007, we put together a financial plan. The sale of our house would net us $100,000; we would pay off the $90,000 FCC mortgage for the land; I would buy a Wood-Mizer wood mill for $5,000 and use it and the remaining $5,000 to build our cabin.

It was simple, but bold. Some of our friends – one couple in particular – politely told us it was a terrible idea. With the best of intentions, they encouraged us to consider all the benefits of a normal house that our little cabin would not have.

"What about laundry?" they asked. "You can't have four kids in a one-bedroom cabin. You'll never get insurance. What about resale value? It wouldn't cost much to make it a bit bigger"

True. All true.

Without a strong objection, they offered, "Why don't you let us draw up a little house plan?"

A few days later, we had nice CAD drawings with many bedrooms and a laundry room. It looked great. But I certainly couldn't construct that very easily with homemade timbers like my originally planned twenty-foot-by-twenty-foot cabin. I did know how to stick-frame. It would be more expensive, but we could use forty or fifty grand to build the house and just not pay off the land mortgage right away.

Spring was drawing close, and we needed to make a call. In hindsight, I suppose I just lacked the confidence – especially after not gaining any confidence from my friends. It did sound a bit smarter to build more of a normal house. We could still build it small. I'd already ordered the Wood-Mizer but decided it would help us with siding and non-structural wood instead of timbers. You're not allowed to use homemade lumber in a "normal" house.

I submitted my house plans and went back and forth a bit until they were approved. It would be thirty feet long and twenty feet wide with a crawl space and a second-level loft contained within an alpine-style, 14/12-pitch roof.

We hired the same old-timer to excavate the footings and, as I explained our plan, he looked at me with disapproval.

"You're not gonna put in a basement?" he asked, cigarette bouncing on his lips.

He wasn't judgemental, but I could feel the pressure to conform.

"It'll just take an extra few hours for me to dig six feet deeper," he reasoned.

It wasn't just the excavation. A basement meant a whole perimeter of concrete walls and stairs and just a lot more expenses beyond the excavation. But I conceded. We had just broken ground and already this project was going in a whole different direction than my cabin-in-the-woods plan from a month ago.

Sarah was old enough to start pounding nails into forms and Rose, having just given birth a couple weeks prior, was carrying five-gallon buckets of concrete around while Abby and Julia played in the dirt piles. We had all sorts of friends and family come out to help through that summer and we got it framed, roofed and locked up.

Our work was interrupted on a number of occasions by elk bugling in the field and, at a certain point, I could bear the temptation no more. I went and shot one.

We clearly had the perfect property and we were just about ready to move in. It seemed we were on the verge of Freedom 27 except for one looming problem. We were seriously overspending. The house itself – the concrete and wood, insulation, and roofing tin – was not the problem, totalling around $25,000. But the well cost over $10,000. The septic field cost over $10,000. And the solar power system was over

$15,000. By the time we were done, we'd pretty near spent the whole hundred thousand surplus from the sale of our first place, leaving us with the entire mortgage outstanding.

I'm not being ungrateful – we had a cozy little house with running water and power and flushing toilets and the most fantastic wilderness/farmland you could want. But we were now solidly indebted and with no clear way out. Maybe, for a while, I just gave up. Over the next two years, we refinanced with FCC, borrowed another $50,000, and built an addition.

I think the truth is, once you've indebted yourself to a point where you can't even imagine getting out, the amount of your debt just doesn't matter anymore. It is a form of giving up. Once you've lost hope, then you just stop caring. I'm not trying to be melodramatic – I'm not talking about being hopeless about life as a whole, just financially. I felt it and, looking around, I see a lot of people who have stopped really caring and just borrow whatever the bank says they can borrow. Debt is a trap. We were in it and it would take some dramatic changes to get us out.

Jeff and Julia framing our new house

Julia helping to clear land on our dream property

Despite the undercurrent of debt, we made huge strides in other aspects of our life on our dream property. Christina was born in the midst of construction and, the next fall, Keziah made an appearance. We didn't set out to have five kids but I do take full responsibility. Like I said earlier, we had a couple before I even knew it was happening.

When Rose was pregnant for the fourth time, she wondered out loud, "Should we stop? Maybe you should get the procedure?"

Aside from the occasional visit, we didn't have any grandparents or aunts and uncles nearby, and four little kids is a lot when you don't get a break. Rose had a superhuman ability to take care of children, keep a house, and keep us all fed but, clearly, we could not have an infinite number of children. We had to find a demarcation point.

We had already sized out of any normal vehicle and Rose and I had promised each other we would die Romeo and Juliet style before buying a minivan. I was topped out of my union pay scale at BC Hydro and, although it was cushy, it was not making us rich. We were both aware of the practical limits as parents, but to just out and threaten a procedure is no way to start a civilized conversation.

"I don't think that's a good idea. I'm twenty-six," I stated to emphasize the absurdity of her suggestion.

Besides, the girls were all so cute and healthy and wonderful. How could we stop? I also considered my own odds of fathering a nation and generally believed more offspring was better. I noticed a mathematical relationship developing between the number of children and the amount of effort required as parents. It was pretty linear from one through three children, with each child requiring about the same amount of effort. But the fourth child required significantly less effort than the previous three – about half, actually. If my analysis was correct, the fourth child was a knee point on a typical saturation curve, meaning the fifth child would require almost no increase in effort.

I believe this to be true for three primary reasons. First, the oldest children are old enough to be quite helpful. In our case, Sarah was often referred to as "little mama Sarah." Second, there is an efficiency in tasks like laundry and food prep where twenty percent more does not lead to a twenty percent increase in workload. And third, at this point, parents just care less – not in a bad way. It's about survival.

"Go for it. Give each other a haircut. Hair grows back."

I countered her procedure offer with a counter offer.

"What if, instead of stopping after four, we have a fifth, but we have it really quickly and get it over with?"

She stared blankly, so I continued.

"It'll kind of be like two for the price of one, like twins without having to birth twins," I was grasping now. "It'll be like a sale, and when you buy something on sale, it's basically making money."

I searched Rose's poker face for signs but came up empty.

"I'll make sure there are no surprises after five," I said, laying down my last card.

Rose was convinced and did not disappoint. She cooked up yet another bright and beautiful daughter. Keziah was born in an ambulance, parked at the clinic in Hudson's Hope.

I had just turned twenty-eight and finally had my complete family. We would do everything together. My number-one goal was to protect, train and raise up five strong, godly women. Each fearfully and wonderfully made, their lifestyle would be a lot different from their contemporaries, but I trusted their unique abilities would make them invaluable to a world in desperate need. When people ask if I'm concerned about the girls not growing up "normal," I try not to laugh audibly.

"The last thing the world needs now is more 'normal' people. It may be more comfortable, but it is far better to have skills, confidence, and a broad perspective. I think it is better to be exceptional."

School is a Joke

When we put the freshly milled spruce board-and-batten siding on the house, it was snow-white but, by the following year, the sun had turned it a golden brown. The colour, reminiscent of a perfectly caramelized campfire treat, led the girls to call it the "marshmallow house." It was a pivotal place in our life in so many ways but one specific conversation during its construction would impact us for decades.

It all started with a chance encounter with the saboteurs. The only part of building I dreaded was drywall. I hate drywall. And not just casually - the flimsy, dusty façade represents everything detestable about modern housing. Anyway, I didn't know how else to finish the house and I was asking around for ideas. Apparently, there were no other options but someone did suggest a contractor who could do a great job. I was told that the crew shows up in a big bus and a handful of the men board, tape, mud and sand, while a handful of women cook three square meals a day. They eat and sleep on the bus and they get it done stat.

The "marshmallow house" with solar panel array

They were the Ludwigs, and they were just over the border in Alberta. I looked them up in the phonebook and made a cold call. Wiebo Ludwig answered and spoke curtly while I explained what I was looking for. He said they were definitely too busy and I thought the conversation was over but I think he heard the kids making noise in the background.

"Do you have a family?" he asked.

"Yep, four kids so far."

His tone instantly transformed into that of a caring grandparent. We carried on with a proper conversation.

"Reminds me of my young family when we were just starting out. We'll find a way to fit you in," he offered.

I thanked him and we planned the date. He stated the price – $7,000 cash. I was new to contracting and that seemed like a lot of money to pay in cash.

I told the guys at work the next day and one of my coworkers responded with, "Wiebo Ludwig? 'The' Wiebo Ludwig?"

I didn't say anything. I was racking my brain, thinking, "'The' Wiebo Ludwig... Should I know Wiebo Ludwig?"

He continued, "Wiebo Ludwig! The saboteurs! He's crazy, and they killed someone! There's a book about it!"

Wow, this was getting exciting. I needed to learn more about the family I'd invited to come work at my house. After work, I went into town to the public library and took out the book "Saboteurs." It's a fascinating story involving a conflict between multinational oil companies and a fervently religious homesteading family. The Ludwigs fought back against harmful and ever-encroaching oil and gas activities, eventually resorting to an escalating campaign of sabotage. It's a dramatic and compelling tale of the little guy fighting back to protect his family.

Of course, some say that Wiebo was nothing more than a crazy lunatic endangering harmless workers. I was in no position to make a judgement but I certainly looked forward to meeting the Ludwigs. Unfortunately, Wiebo and the big bus never came but they sent a satellite crew – just a couple of the boys – from the mothership that was working near Chetwynd at the time. They worked hard and did a fantastic job on the drywall, and I remember distinctly when the wind ripped seventy one-hundred-dollar bills out of my hands and all of us scrambled across the yard to collect them.

The most memorable, however, were our wide-ranging discussions at lunch. The oldest son was fairly bold in conversation and asked about our life and our faith, and we eventually came to discuss the girls and their education. Sarah was five and had just started grade one at the public school in town. Rose and I weren't particularly happy about it but what else could we do? When I was growing up, homeschooling was pretty weird. My cousins did it and they were super smart and healthy, which was weird. Other than that, I had always imagined all the other homeschoolers had fanatical parents or were too ugly to go to school.

In Hudson's Hope, however, we had met a number of lovely families who homeschooled their kids for religious, academic or just practical reasons. We tried it out the year before by homeschooling Sarah in kindergarten – which is kind of a joke because kindergarten is all about snacks and storytime, which is exactly what a mom would already do for her four-year-old.

I remember going to preschool, which was like jail for children before they were old enough to go to kindergarten jail. I don't remember much of the preschool but I do remember running away. I rode my tricycle to preschool that morning and I used my tricycle to get away. Logically, I chose to ride away in the downhill direction. About a block away, just as I was gaining speed toward a busy intersection, a helpful lady ran out of her house and physically stopped me.

Back to jail.

The kindergarten experiment worked fine but Rose and I could not shake the idea that homeschool was weird and normal school was good. It also seemed like a lot of work and Rose had a lot of work to do already. Yet, sending Sarah to school had practical problems as well. It was a half-hour drive each way if we were going to drop her off, and if she took the bus, it would take forever – approximately an hour and fifteen minutes each way. I had commuted for hours a day back at BCIT and I couldn't stand the thought of wasting Sarah's time like that.

More importantly, we just had a bad gut feeling about giving up Sarah at such a young age. She would be gone nine hours a day and she would sleep for nine hours a night, leaving six hours a day awake at home. At the age of five, her parents would become a minority influence in her life, the majority being teachers and other students. I know lots of teachers are good but I've certainly had some bad ones. And in a classroom with thirty kids, well, it's actually the students who are in charge. With five kids, it was nearly impossible for Rose and me to keep on top of them, correcting bad behavior and encouraging good. How much "parenting" could a teacher do with thirty kids?

So we would throw our precious young child – barely more than a toddler – into a giant building full of people and hope that her teacher was good and all the students around her were good. I knew children had to grow up and, although they start one hundred percent dependent on their parents, they end up totally independent. I knew she would need to experience the influence of other people but it seemed like too much too fast.

Despite our emotional hesitation, we did the normal thing and enrolled her in school. We didn't feel great about it but she seemed to be doing okay through the first few weeks. I told that all to Ludwig Jr.

He replied, "So if you think it's best for her to be at home, why is she at school?"

His question was a direct hit, and I half-heartedly offered the lame excuse, "We don't know how to homeschool."

To this, he went on to tell the story about how the kids in his family were homeschooled.

He said, "The girls tend to love stories and are eager to learn to read on their own by the time they're five-ish. As a parent, you just need to be a little intentional about helping them when they're ready. Boys, on the other hand, don't want to sit still, and so they let the boys tag along and just help out doing stuff. They still love stories – by the time they're seven or so, they can pick up reading in just a few weeks. You could try and force them and you'd just have a fight on your hands, or you could let them learn it when they're ready."

He described how older kids learned real practical skills like gardening, food preparation, fixing trucks and building fences. At the end of it all, he challenged me with a simple statement.

"Maybe you should do what you think is right and figure out the details as you go, instead of doing something you don't believe in."

It felt like righteous conviction and I couldn't argue. In fact, it was a profound idea that had implications far beyond our current homeschool decision.

"You're right. . . I think you're right," I admitted. Rose also felt total peace and we notified the school. When we went down to meet with the principal and pick up Sarah, the teacher came out to meet us, dressed as a witch. For real? It eliminated any remaining doubts we had. It was mid-October, the classroom was decorated with all sorts of wicked and death-related Halloween crap and the teacher was dressed like a freaking witch.

It perfectly reflected my complaints: How is any of this time and effort helping Sarah to learn reading, writing, and arithmetic? It seems like a lot of filler at best and indoctrination at worst. And what kind of philosophical and spiritual influence is this teacher trying to have on my child? How can I instill core values into my children when I immerse them in the totally unknown values of strangers?

We felt terrific about our decision and, in the next weeks and months, we came to understand the massive scam that is elementary school. With all our girls, meeting government-established academic achievements took less than an hour a day. Not to get all mathematical but if the actual learning takes one hour, and a normal school day is about seven hours, it means that regular school is about fifteen percent efficient. If your car was fifteen percent efficient, you'd push it off a cliff. If you were fifteen percent efficient at work, you'd get fired. If this book had fifteen percent written pages and eighty-five percent empty pages, you'd demand a refund. If you truly evaluated the school system for its learning outcomes versus cost, you would discover a monumental waste of time and money. Only when you evaluate it as a babysitting service for working parents does it make any sense.

At first, we took it one year at a time. Each new year we happily decided to continue homeschooling. We weren't sure how it would look in the high school years but we would cross that bridge when we got there.

By far the most common question we would get from other curious – and sometimes judgemental – parents was, "Don't you worry about your kids' social skills?" or "How do you make sure they learn good social skills?"

The first couple of years, I didn't really know what to say, but I'd reply with something like, "Oh, we make sure they get to spend lots of time with friends and other kids. . ." But now I just chuckle and answer with my own question: "Do you think school is good at teaching social skills?"

I think school is terrible at teaching social skills. It's not one of the prescribed learning outcomes and, even if it was, how could one teacher provide meaningful advice on the social interaction of your child amongst all the other children? In reality, kids develop coping mechanisms to deal with the mob. They form small cliques. They sometimes learn attention-getting tools like bullying or making fun of others. Alternatively, they learn to withdraw. No, kids should not be learning social skills from a mob of other kids.

It's also a very strange social scenario that doesn't exist anywhere in society outside of a school. Where else does a person find themselves amongst thirty others, put together only by their production date? That doesn't sound like any workplace or even college or university, where people gather with similar interests and are not segregated by age.

Our kids never grew up trying to cope within a school setting. Instead, they completed academic studies, developed skills and learned to relate to people of all ages and in a variety of settings.

After one month in public school, Sarah completed eleven-and-a-half years of real-life education and her fair share of annoying, government-prescribed assignments. As I write this, Christina and Keziah have just completed grade twelve as well. Rose and I have overseen sixty student years' worth of K-12 education, all graduating with BC Ministry of Education Dogwood Diplomas.

I can now concisely summarize my sentiments. And I'm not being flippant when I say this, but I think the path forward for our society is to completely remove government from education. They are now obviously prioritizing the system for babysitting versus educating. Even if they did keep to education, they're not good at it. Why would they be?

They're OUR kids. As parents, we care the most about our kids. We want what's best for them. I'm not saying we are necessarily the best to teach a certain subject, but we are best to direct their education. Government workers obey government policy, they are limited by government budgets and influenced by government ideology. They do not effectively care about our children. Not like we do.

Furthermore, most of what is taught is not retained because it's never used. It's not practical knowledge. The idea of a "well-rounded" education sounds good but, in my experience, it's more of an academic vanity project. Wouldn't it make sense to focus on core academic skills: reading, writing, math and applied science? How about letting kids pursue their interests and learn skills from their parents for the "well-rounded" bit!?

A normal kid spends about thirty-eight percent of their conscious life at school. Is the government-mandated academic curriculum that important? Is the same curriculum mandated for every single child in the country truly appropriate for your child?

Once you start questioning the status quo, there's no end. Eventually, it becomes easier to imagine a world without government education and, then, consider how you can best prepare your kids for the future. For some kids (maybe most), academics shouldn't be the primary goal. Character traits like hard work, integrity, ingenuity and interpersonal skills are not academic but are universally valuable. As parents, our primary concern should be growing our kids into good humans and, yet, their entire development has been hijacked by academic bureaucrats.

Take back control. Think outside the box. Question the norm and raise great children!

Off the Grid

Going off-grid in 2007 on our dream property was, firstly, a financial necessity and secondly, a fun technical challenge for me. We got a quote to install power lines to the house and it would have been at least $80,000. Additionally, we'd have to look at an ugly power line in front of our house for the rest of our lives.

It was probably our friends, Dwayne and Julie Biever, who had raised their kids and lived off-grid for decades that gave me the confidence. They had a beautiful place – south-facing, overlooking a lake, big luscious gardens, a trout pond and direct access to the backcountry. They had carved out their own little piece of paradise thirty years prior, and it was an inspiration. In those days, there were no online stores selling cheap solar power equipment and even getting information was tricky.

I found a small company on Vancouver Island, Energy Alternatives, and got on the phone. They could definitely sell me everything I needed, but it was not super cheap. I remember being excited that solar panels had recently come down in price from about eight dollars a watt to seven. We settled on three 175W panels, a Xantrex 4,000W inverter/charger, and a large twenty-four volt lead-acid battery bank. The whole thing cost about $12,000 and was sufficient to run a well pump, a welder, and anything in Rose's kitchen.

We also bought a really nice 4,000W Honda generator, which cost another $4,000. I built a stand-alone panel rack from an old salvaged satellite dish mount. The whole system was a fraction of the cost of getting grid power to the house but left us with a few limitations.

Most importantly, it did not have enough energy storage to allow us to use a block heater, which was a big deal. I had not thought of this ahead of time and, with the first arctic blast of fall, it hit minus -30°C and left me in a jam. At the time, we had a '97 Ford F-250 with the big-block 460 running on gas and propane. I phoned around and learned that the old-timer trick was to lay a tiger torch in a big cast iron pipe on the ground underneath the front of the truck. It'd get red-hot and radiate heat up to the engine and, in forty-five minutes – presto – your engine turns over and purrs like a kitten.

I did not have a cast iron pipe, but I did have a short length of single-wall stovepipe, which I promptly put into service. And boy, did that make me nervous. I tried my best to adjust the tiger torch, but it kept wanting to spit flames out the other end, curling up and licking the back of the oil pan. Even once I had it adjusted properly with no visible flames, I would go back in the house and just stare at it from the front window. Plumes of steam pouring out in all directions; it looked like a spaceship ready to take off. I was warned that, while the torch method worked, it also burnt down trucks consistently. It was an anxious ritual that I was determined to eliminate.

I really internalized that particular challenge of off-grid living and, the following year as part of our house addition, we added an insulated below-ground garage. Even unheated, the surrounding earth moderated the temperature so much it rarely hit -10°C. And that was a really important lesson to learn regarding extreme temperature swings. Insulation and thermal mass can passively eliminate the peak highs and lows and avoid the need to heat or cool a space. If the average temperature is acceptable for your purpose, then you can just build in the ground and that's all you need.

The epitome of this principle, of course, is the root cellar. We never did get to build a root cellar at the marshmallow house, but we would eventually have one and it would be fantastic.

We bought Rose a propane stove and ran it off twenty-pounders. It would seem natural to plumb in a propane hot water tank at the same time, but we could only afford the essentials and hot water wasn't essential. Rose had a fridge, an oven and power in her kitchen. We had running water in the kitchen and the bathroom. Aside from things like the block heater issue and a few other quirks, it hardly even felt off-grid – until you had a cold shower.

I specifically remember the first time having a shower with our new water system. We didn't have a shower curtain up yet – it was a tub-shower combo – so I contorted my body to minimize back-splash while also prancing around to keep any one body part from freezing solid. It was shocking, to say the least, but still very rewarding to know that this water came out of our own land, through our own well and through all the systems that I had bought and installed myself. All that satisfaction kept me from really acknowledging the nerves sending "icy pain" signals to my brain.

It was with diminished reasoning that I contemplated how to get my head wet and wash my hair. There was no feasible position to prevent water from splashing everywhere. So, I turned off the shower head to allow the water to jet out of the tub faucet, knelt down, and stuck my head into the flow. Up until this point, the whole experience could have been described as brisk or refreshing, but it quickly changed. I've bathed in icy waters, but I suppose I've never held my head underwater in a swiftly-moving icy river. Totally unsuspecting, I afflicted myself with an externally-induced brain freeze of epic proportions. I didn't even know that was possible.

At first, everything just went numb, and I scratched at my scalp with my fingertips and my head felt nothing. It was like my face was still there, but the rest of my head had just fallen off. I wondered if this was what drugs were like. Maybe I was overdosing? With a stroke of genius I turned off the water, and only then did the pain start. Slowly at first, it built up in classic brain-freeze style, leading me to howl and jump and slap the top of my head all at once.

"Rose, Rose...ROSE!!" I yelled.

She ran down the stairs and burst into the bathroom asking, "What? What's the problem?"

She came for the emergency but stayed to watch me dance.

"I'm fine now," I assured her. "The water is actually fine. You can have a shower now if you'd like."

With an eye roll, a smirk and a, "Nope! I'm good," she whirled out and left me to thaw.

Having proven its viability, we nonetheless generally chose to heat up a pot of water on the stove and pour a bath, reserving showers for guest use only.

The hunting was phenomenal and we had success shooting big whitetail bucks, bull elk, mule deer and black bears on or near the property. Rose and I would sometimes go on hunting dates and sit at a little tree stand in the back corner in a mature aspen grove, watching elk and moose and deer – and sometimes shooting one.

My brother, Shawn, came up and shot a seventeen-point non-typical whitetail – a beauty he would put on the wall. We had always hunted our red meat and we'd had thin years before, but we were now amidst a season of plenty. As hunters know, sometimes you get one and sometimes you don't. We had gratefully accepted from others when they had a surplus and were always happy to share when we tagged out. An integral fabric of primitive hunter-gatherer communities, sharing in a successful hunt is still a big part of being a hunter. Even then, when the girls were young, we were a three-deer-or-one-moose-a-year family.

Besides hunting, we managed a terrific garden with help from the pigs, who cleared the ground for us.

I protected those pigs from a giant black bear in early spring. Our new German Shepherd pup, Hunter, raised the alarm and I ran to the front door to see a big, black sheet of plywood step on the top wire of the pigpen and stretch it effortlessly to the ground. Within seconds, the bear was lurching around the pen, suffering from decision paralysis with eight squealing wieners running in all directions. I quickly retrieved my 7mm and dropped him right there in the pen. That was the biggest black bear I've ever shot and I needed help from a friend to drag him out of there. He must've stuffed himself with oats all fall because, even now, just out of a winter's hibernation, he had a three-inch rind of fat on his rump.

Jeff, Rose, Sarah, Abigail and Julia with an elk on our own field

Jeff and Sarah with a beauty whitetail

I had grown up taking meat to a meat cutter but it had often been hard to find someone with availability during peak hunting season. And the quality was sometimes questionable. It was Dwayne Biever who asked me, a little bewildered, why we didn't cut our own meat. I had no good answer. I guess I just didn't know how. Remember, I'd only just escaped the suburbs a few years ago.

Dwayne gave me these fabulous instructions for cutting our own meat. Two rules. One: "If it looks like it comes apart, then take it apart." And two: "If it doesn't look good to eat, then cut it out." We've been cutting all our own wild game and domestic animals ever since and those two rules have not led us astray.

Cherryville Interlude

Just like the time spent on our first five acres, we worked on the dream property non-stop – building gardens, pastures, wood sheds, tree stands, and all manner of small home improvements. And, similarly, we would leave this one after just two years. It was our dream property and it was fast developing into a beautiful, comfortable, productive homestead. Our house was built directly in the middle of one of the best hunting properties on the planet. We had made good friends and we felt at home.

There was only one problem, and it was me.

Maybe because I'd missed my Freedom 25 plan and just sandbagged my Freedom 30 plan, I felt the need to shake it up. Or maybe those subconscious thoughts were just contributing factors. Either way, my day job was killing me. Work had stopped being fun. I didn't feel technically challenged. There were no leaders I was inspired to learn from and the unionized atmosphere was not the strive-for-excellence team I was hoping to be part of. There were plenty of people I enjoyed working with and many that were admirable and good friends. But we were not pulled together in a team that pushed ourselves to be better and achieve great things.

The us-and-them, unionized, government bureaucracy could not be my lifelong home. But I wasn't ready to leave the company yet and there was a very interesting opportunity I thought I might be good at. It was a two-year, temporary position for a technical training instructor. They would sponsor me for a diploma in adult education, and the pay – fifty-one dollars per hour – was the highest in the union.

I liked the idea of learning the secrets of education – after all, I was the principal of Off-Grid Homeschool – and the pay could only help. The only hang-up was the location. Based in Vernon, we would need to move. If it had been a full-time job posting, we never would've done it. We could never have walked away from our dream property. But the company was going to pay a stipend to help us with housing in Vernon and we could keep our property and rent it out while we were gone.

Jesus said, "He who loses his life will find it." Similarly, we were about to lose a ton of money but find Freedom 35.

Vernon is a nice little town, but it is hot, and it is a town, so living there was not an option. We scoured the rural areas in all directions and found Cherryville, about forty-five minutes to the east. It had one little gas station and a general store called Frank's on the main highway. Like a typical small town general store, it even had locally-made arts and crafts. Near the till, there was a little stand with handmade postcards. One of them read, "Cherryville – not so much a place, but a state of mind."

That summed it up pretty well – an eclectic mix of loggers and hippies, nestled on the edge of the Monashee Mountains with the ice-cold and picturesque Shuswap River flowing through it. Not a cherry tree in sight. Apparently, the Cherryville name was more of a marketing slogan than a description.

We went to the little church on Sunday morning and chatted with a lot of friendly folks after the service. Specifically, there were two families, each with kids similar in age to our own. Eric and Lisa had two little girls Sarah and Abby's age, and Dave and Shoshanna had two little boys Julia and Keziah's age. They were so friendly and inviting. We were immediately convinced that Cherryville would be our new home.

We had the audacity to ask Dave and Shoshanna if we could camp at their place for a few weeks while we found our own and they had the grace to let us. We had moved around a bit before and sometimes making new friends took months. But this felt more like we parachuted right into a friendship convention and got straight to it.

We looked for places to rent and found very little, but the 2009 housing market was pretty soft and I figured we should probably just buy something. After all, I was already $150,000 in debt – why not add some more? The crazy bank lady said "No problem," and loaned us $350,000 to buy a house on ten acres. We were officially land barons.

Our new place had a mountain view and a cedar forest hopping with whitetails. And, like a throwback to old life, it was grid-connected. We would have electricity but, as it turned out, we would not have much water. The shallow well lasted only a couple weeks before running dry and we would go on to spend $45,000 drilling holes. It was just the first expensive lesson we would learn.

The instructor's job was fantastic. I had a great boss who lived and worked eight hundred kilometres away and gave me about that much latitude to develop and deliver my training courses. I enrolled in Vancouver Community College's (VCC) Professional Instructor Diploma Program and immediately gained insights into the ways people learn – and, maybe more importantly, the ways they don't learn. I was able to steal and scavenge various pieces of equipment from around the BC Hydro system and put together a small-scale, hands-on, functional training facility that enabled us to model complex generator systems and transmission line protection and control systems.

For the first six months, I developed forty hours' worth of instructional material to be delivered to a technical audience. For the next six months, I delivered the course to engineers and technologists. I repeated the same pattern in the second year. I took the odd week off to complete a course at VCC or sometimes completed them during consecutive weekends. Work was challenging, fulfilling and rewarding. And thanks to another new friend and fellow instructor, Gary Davidson, it also elevated my ping-pong game to near Forrest Gump levels.

We tore into our new ten acres just like we had our last two properties and quickly turned up a garden, renovated the chicken coop and added some birds. We fenced off the front acre for a couple of hair sheep and goats to keep the scrub down.

We built a ground blind and, eventually, a tree stand at the top of the property and spent countless evenings up there watching the whitetails. Sarah was eight and Abigail was six, and this was the perfect hunting training ground for them. We almost always saw deer and the girls were able to observe just how powerful their eyesight, hearing, and smell were. Just by shifting their weight, deer in the clearing would stare directly at them. A crinkle of their coat and a deer would jump to attention, turn its ears at them accusingly and then tear off. They learned very quickly the incredible stealth required to go undetected.

The learning curve included all the practical elements as well. During those years the girls worked with us on many successful hunts, helping gut and skin the carcass and cut and wrap the meat. My new friend,

Dave, joined the fun as we helped him shoot his first deer with archery equipment.

We were not used to living close to neighbours and, with another ten-acre property next door, it felt like we were practically roommates. They were kind and it was never an issue, but their dog was a bit of a womanizer and our dog, Hunter, was available. She was a purely outside dog, a bear dog, and, back in Hudson Hope, she just did her circuit around the house in the yard and never ran away, roamed, or otherwise got into trouble. But now she was in a neighbourhood and we didn't know what to do with her. She was coming into heat and we were torn. We didn't want her bred and we also didn't want her lured out by the wolves or coyotes.

"We could keep her closed in the garage at night," I suggested. "But she can't chase off bears from in there."

"I know," Rose added. "And I just feel bad for her stuck in there all night. I could just leash her on the porch."

The neighbour's dog had come around the last few days, but Rose had just shooed him away. One day, he came by and could not be dissuaded and Rose had to encourage him with the business end of a broom. That night, just as we were falling asleep, Rose heard Hunter make an unusual sound, like a howl but almost more of a moan.

"Oh no!" she said, and bolted out of bed. "That's that dog!"

I was getting up, groggy, but she was already running to the front door. To add value, I called out to her, "Be careful!"

She grabbed the broom on the way by and threw open the door. Finding the dogs stuck together, she let out a furious battle cry.

"Get. OFF. Of her, you...." She charged at the guilty couple in her underwear, yelling, "Get! OFFFF!"

I got to the door just in time to see Rose step on the ice, lock up, and then throw both feet forward in flying ninja style.

"Wow," I thought out loud, "She's gonna dropkick that dog."

Rose just kept flying through the air, down the steps, off the porch, still clutching the broom, and then hit the gravel driveway. Her "AAaaaahhhhh" through the air turned to an "Ooowwww" as she slid on the gravel. She was MAD. And it was all the dog's fault. She jumped up swinging and yelling.

"Get, get, GET!!" and whacked the soon-to-be-father with the broom.

He yowled and popped off poor Hunter, but it was clearly too late. Puppies were in the mail.

It wasn't just the dogs. It turns out a six-month-old male goat can also make it happen. And the thing about billies, they make such a big show out of it. A mere month earlier, Junior was a cute, cuddly, bouncy, fluffy, adorable little baby. But like a gremlin, overnight he had turned into a lip-curling, greasy, pee-soaked, grunting, posturing deviant. We immediately had to implement a no-petting-Junior rule. We hadn't planned it this way but we also immediately commenced Biology and Reproduction 101. Not a lot of romance to sort out – just the hard facts. Of course, it's best this way. Why read a textbook with all its awkward descriptions when you can just grab some popcorn and watch the show? The family farm has sex-ed beat hands down.

We continued to develop our homesteading skills on the edge of the Okanagan desert, like a prolonged working holiday. Rose was homeschooling three kids and protecting the world from the youngest two. She was busy but content with her kids, her home and her garden. I was fully engaged at work and loved having a new property to play with.

We were getting to know our new friends, the Larsons and the Godbers, and we'd often go down to the river on summer evenings and play pond hockey in the winter. I bought a little 600cc Suzuki S40 motorbike to commute with and take the girls on quick joyrides in the summer. We collected Grandad's old apple juice equipment – the grinder, the press, the settling tanks and the brew pot with a spigot – and my extended family came up for a weekend in the fall to make apple juice.

For a holiday, it was fantastic and we loved every minute of it. We considered settling in for the long term by checking out some land boasting big mature cedars and little mountain streams. I was dying to set up a micro-hydro, off-grid house on one of them. There was a possibility this teaching job would turn full-time.

Dave and Shoshanna were looking to buy their first place as well but, like us, found it to be very expensive anywhere within an hour of the vineyard-rich Okanagan Valley. That was an issue but I also couldn't imagine myself settling down in a place so tame. I mean, we had deer in our yard and that was wonderful, but I wanted a yard where we'd have a decent chance of watching a grizzly bear chase down a moose and eat it. I wanted to live in a place where I could walk out my back door, head north, and hit the Yukon border without bumping into anyone. Dave and Shoshanna were open to moving out of the area and we started looking for places together.

Going right back to our land in Hudson's Hope was the obvious option for us. Our place was perfect, except for the mortgage. We had friends up there but at this point in my career, I was at a crossroads. After nine years at BC Hydro, I knew I couldn't stay there much longer. I didn't feel ready but I knew the time would come when I would have to forge a path of my own. The problem with Hudson's Hope was that the dam was the only show in town. The BC Hydro job board had a posting for a sub-foreman serving North Central BC. The job was interesting because I'd never held a permanent crew-lead position, and it would also put me back into transmission and distribution, which I was a little rusty on. If I wanted to round out my skill set for future contract work, this would be valuable experience. I also knew the boss and he was someone I could learn from.

It did seem a little risky. I had two separate properties 1,000 kilometres apart, each with their own mortgage. Now I was applying for a job nowhere near either of them. But Rose was happy with the plan and Dave and Shoshanna were excited to look for land together with us. If we found the right piece, we could buy it together and split it. I took the job and put the Cherryville property up for sale.

Warning: tough lessons ahead. . .

A Financial Crisis to Call Our Own

I was never interested in land in the way a realtor or investor might be. I just loved the possibility of dirt and trees and rocks and creeks. I knew land was generally a good investment, but I was not an investor. I had paid attention to the world's financial events around 2008 and, although Canada was not hit as hard as many nations, there were impacts. When we bought the place in Cherryville in 2009, I figured prices had dropped and were stable but would inevitably go up. It turns out I was absolutely correct, except I didn't have a firm grasp on the "when." If I could've held that property for ten years, it would've tripled in value, but I'd backed myself into a corner and was forced to sell now.

I had managed to put the combined $500,000 mortgage out of my mind for a while but now it hung like a concrete block around my neck. I had to sell it or there was a distinct possibility we would default. We might be able to rent it and squeeze by but we might not. Renters have the potential to ransack a property and refuse to pay. We could be screwed.

We already lost $45,000 on the thousand-foot stone holes we'd drilled and maxed out a line of credit to pay for it. We listed the house and, just as we feared, nothing happened. After a couple of weeks, we finally had our first showing. After letting us sweat it for a week or so, they made a low, but reasonable, offer. It came in $30,000 lower than the price we had paid two years earlier.

Ouch. I was feeling kinda like a dunce. I had backed us into a financial corner. We hummed and hawed but accepted the offer. It was subject to financing and a home inspection, so we had to wait a few weeks before we had certainty. In the meantime, I talked to the banker about us not getting enough money from the sale to pay off the mortgage. She seemed pleased.

"No problem, we can just refinance your other place."

"Great," I thought. "Why don't you just pour concrete blocks on my feet while we're at it?"

That's when she told me about the mortgage penalty. Penalty? Why would I pay a penalty for paying off a mortgage? And then we looked through the fine print and found the little equation that would cost us another $15,000 just to pay off the mortgage. Apparently, we had locked in for five years.

I wasn't feeling very good about the state of things and when we heard back from our realtor, it got worse. The inspection turned up some moisture inside the sealed window units and windows are super-expensive to replace, so the buyers lowered their offer by $12,500.

I felt like a kid who disobeyed my mom.

"Don't eat these cookies, Jeff. I'm saving them for tonight's company."

But the cookies looked good and they smelled good. So I ate one, then another, until the jar was empty. The result? A sick feeling, then shame when discovered, a spank on the bum and the upchuck of those precious cookies.

I'd been caught with my hand in the cookie jar, and now I was feeling shame. By the time we paid for the well, the realtor commission, the penalty, and the loss on the house, we were down about $100,000.

We made a trip back to Hudson's Hope and madly spruced up the house and yard in time for an assessment. Thankfully, we still had equity remaining and we refinanced to absorb all our losses. Unfortunately, our mortgage was going the wrong way and was now sitting around $250,000. Our escapade in the Okanagan turned out to be very expensive. It was a huge setback to our financial plan. But newly-developed skills, lifelong friends, and life experience are hard to put a price on. And in a few years, that extra debt would push us to make a very difficult, but life-changing, decision.

With the financial crisis averted, we were able to turn our attention to the adventure in front of us. I was starting a new job and we were moving onto some new land. We discovered a lonely little forty acres out in the middle of nowhere. It was unassuming and cheap. We went to look at it with Dave and Shoshanna and we all fell in love with it. Sharing it with them would make it even cheaper. Good thing, too! We were tapped.

We had fallen into the debt trap, even though we were overtly aware of it. We had learned our lesson but, embarrassingly, we still needed to use a line of credit to help us pay for our half of the land. This time, things would be decidedly different. We were getting a second shot at our original goal and we knew exactly what to do.

It was time to head to our new land and start all over again.

To most people, freedom is the ability to do whatever you want. This definition fits with the Freedom 55 plan I remember being advertised as a kid. The key was having enough money – enough money to pay all your bills and even more money to do what you wanted in all your spare time. For the most part, I think this is what most westerners are still aiming for but, on deeper inspection, it's a rather weak definition of freedom.

What, exactly, are your needs? And how much money is required to do whatever you want? There's always going to be something you want to do that's too expensive, no matter how big your savings are. It's only a select few who ever have enough money to do whatever they want.

But this isn't the biggest problem. The TV commercials always showed a grey-haired couple eating at a poolside fine restaurant, golfing, or laying on a beach somewhere. In short, you work a job you don't enjoy for thirty to forty years so that, one day, you can quit cold turkey and live a life of leisure and entertainment.

Some might be indignant that I presume you don't enjoy your job. Fair. But why doesn't the commercial show you still enjoying your job? It's clear from the marketing that we imagine it is a worthy goal to pursue a life of self-satisfaction rather than a life of fulfilling work. But is it? Should we work towards a goal of not working? Would we really find a life of leisure fulfilling?

Let me loop back. I understand the majority of us make the best of our employment. We enjoy it for the most part, appreciate most of the people and feel relatively productive. OK, but if I gave you ten million dollars right now, would you quit? I bet you ninety-nine percent would. And there you have it – ninety-nine percent are just putting up with our jobs, while only one percent actually love it.

We often get so focused on monetary compensation, we forget about the purpose and productivity work provides. We forget that it's a way for us to contribute to our families and the community. I admit that early on I adopted the common, shallow view of work and missed its deeper value.

What if, instead of chasing the money so we could stop working, we spent thirty to forty years pursuing work that we loved and felt passionate about? What if freedom was not freedom from work, but actually included productive and rewarding work?

Another red flag about a traditional financial retirement plan is that it's not traditional at all, in historical terms. Only in the mid-twentieth century did retirement gain popularity, resulting in billions of dollars of profit for financial institutions. It's always good to be skeptical of advice when the advisor stands to make money.

At some point in my early thirties, I came to view retirement as a wildly successful marketing product that left retirees with an unfulfilling life of self-gratification and an insufficient lump of savings.

My retirement plan would look totally different. I would pursue skills and work that I found satisfying and, in turn, would make myself productive. This would result in a better product or service to society and fulfill my very human need for purpose. Instead of investing in registered retirement savings plans, I would invest in a debt-free homestead that produced its own vegetables, bacon, and fresh milk. It was a liberating revelation, immediately allowing me to view my work as part of the goal instead of part of the problem. Instead of striving for a seemingly insurmountable financial goal decades in the future, I could start living the dream right now.

I could dream of a time in twenty or thirty years when I would be a well-respected businessman providing good products with integrity; a legendary, old, wise and passionate teacher; and a grandfather tending a vibrant homestead with the help of a small army of grandchildren. I didn't imagine doing these things full-time, slaving away. I wouldn't need to. If I could remain relevant and productive, I would continue to earn well into "retirement."

When I considered the work I wanted to do, I never presumed it to be forty hours a week, Monday to Friday, all year long with a couple weeks' vacation. Instead, I could work intense hours on a big project and then take half the year off to hunt and fish and garden. Or I could chip away twenty hours a week and spend the rest of the time on the homestead. Products and services come in all shapes and sizes – short-term and long-term, one-time, single events, and evergreen. Once out of the traditional career mindset and into the self-employment and entrepreneurial realm, the world turns into a palette of opportunities.

And so, back to the definition of freedom: it is not freedom from work, and it does not require hoards of cash. Fundamentally, of course, freedom is having autonomy over your words and actions in all of life's choices. The ability to make our own choices does not relieve us of the consequences of our actions or from the laws of physics, for that matter. Freedom does not mean that we get to do whatever we want. I want to fly, but the tyranny of gravity won't let me! If I make poor choices then I have also chosen the consequences of those actions. If I want to eat, I need to work.

This does not mean I'm a prisoner; it means I'm human.

To this extent, work has been a curse on humanity from the very beginning. We've had to produce our food by the sweat of our brow. But, examining my own work history and observing those around me, it seems like we are all working a lot more than required for just food. I have seen many people working long hours and spending a lot of time driving to and from work and not really enjoying it. They should at least be eating like kings, but often consume fast food. It seems out of balance – almost like most of our labour and the majority of our money is being misdirected to less important things.

Our houses are large and some have things like crown moulding and wainscoting. A lot of the house dormers are facades only – there is no liveable space in the attic. Most complex rooflines are exclusively for looks. None of that's a huge deal, except that we're not just paying for the house and its frills; the payments we're making are mostly interest on those frills.

We have multiple vehicles per household, mostly so we can commute to work. And then there is the cost of going to and from work. We pay for insurance. We pay for fuel. We might even pay for daycare while we're at work.

Then there are the things we do for fun – drink fancy coffees, pay for movies and TV shows and generally buy extra stuff. The extra junk is not a crime, but when you add it all up and look at the numbers, only a very small percentage of our labour converts to nutritious food and warm, dry shelter. Think of it like this: we work on Monday for a warm, dry house and nutritious food and the rest of the week for all the fluff.

Is that what we want?

Some have never thought of this lifestyle as optional – they're just doing what everyone else is doing. Some feel trapped, but they don't know how to get out. They feel the system is making them do it and taking away their freedom. And it clearly is.

Why can't you just buy a small lot and build a small, simple house?

Why can't you do what my newlywed grandparents did – renovate an old chicken coop and live in it for a couple of years while you build up your skills and save up your money?

Why?

Because in the last seventy years, we've added a thousand laws, rules, regulations, policies and ordinances. Yes, each one of those well-meaning restrictions on your freedoms is for your own good and to keep you safe. The end result is death by a thousand cuts.

Without getting too political, let me say that restricting personal freedoms in exchange for government control might work in theory but is never beneficial in real life. The great achievements and wealth in North America stemmed from the uninhibited productivity of free men and the stagnation we see now is the result of central planning. There is much we can do to diminish our governments and empower our people, but that is not the purpose of this book. For some, their life's purpose is to fight the good fight. For others, it's to simply find a way through for their family. For the rest of us, it's a bit of both.

There are certainly many ways you can buck the system and still live within its cities but, for me, there was only one way.

Get out.

I knew when I was nineteen that if I chose to stay in the city, I would not be able to achieve my kind of freedom. I would not be debt-free. I would not be free to build my own house. Most municipalities won't even let you have a chicken coop, let alone live in one.

I would not be free to pee on the bushes in my back yard. I'm sure they have some sort of rule against that.

I would not be free to grow my own fruits and vegetables or raise my own bacon. You have to be wealthy to have enough land for that in a city.

Why is it unreasonable to save up for a couple of years and buy a house outright – for cash? Why is it crazy to think that I shouldn't need a massive, multi-decade debt just to get a little shelter?

Take a drive through an average subdivision in an average Canadian town and almost every house you see is being paid for at least twice – once for the sale amount and then more than that amount again for the interest. Add a hefty property tax and you really have no choice but to settle into the rat race for the rest of your life.

The city can be a bit of a black hole – it's packed with people who believe that's where they want to be. But most have never given another lifestyle a chance. We got out of the city and never looked back. We didn't go straight from the suburbs to an off-grid enclave in the woods – it was a process – but we were always headed in the right direction. We didn't immediately free ourselves from the system, but each major decision allowed us to breathe easier and see the limitations fade away.

Our freedom plan was taking shape and maturing. Freedom 25 was about building our own house debt-free. Interrupted by an onslaught of precious little daughters, we extended the deadline and pivoted to a Freedom 35 plan. We would own a debt-free homestead with productive gardens and fencing for domestic animals, surrounded by the beauty and provision of nature. And we would continue to work – both the passionate and productive income-earning kind and the holistic, ancient kind that included hunting, gathering and gardening.

We had a plan, but could we pull it off?

The Hovel

It was early summer 2011 and, having put "normal" to rest months ago, we fully committed to the primitive task of building our "Tarp House". Fighting through the mud and mosquitos, we simply addressed the most urgent needs as they arose. A bathroom, a roof, water . . . We were carving a life out of raw land.

The tarp house at Tipper Creek was a thing of beauty and those first few weeks in our homemade hovel were glorious. It was early June and the sunrise at 4:00 AM was followed by a symphony of songbirds. With no insulation and much of our walls just covered with screen, the nature surrounding us was immersive. We heard the birds and the wind and the wolves. We felt the heat and the cold and the damp and the breeze. We were living in a giant tent in the middle of the woods and it felt great.

With almost seventeen hours of sun per day, there was an explosion of green. Dark green, light green, high green, low green. In just a matter of a couple days all the aspens morphed from dormant white skeletons into bushy-topped emerald popsicles. The grass sprang up alongside the wildflowers. The forest floor and any natural clearing was wall-to-wall with red Indian paintbrush, fancy little columbines and wave upon wave of purple lupines. Delicate little white flowers gave away the location of all the saskatoon berry bushes.

We were ecstatic. We had accidentally bought a magical park of wildflowers and songbirds.

We needed help to occupy the girls while we worked. Local friends were in the future and grandparents and aunts and uncles were a thousand kilometers away. So Rose found an ad in the classifieds and, next time she went to town, stopped by to pick up a kitten. We knew we would need a farm cat to help us with the mice, but it occurred to us we also needed a kitten to help entertain the kids. Of course, Rose brought home three kittens because, hey, they were kittens. I figured if they could survive the love and attention our girls gave them as kittens, they might stand a chance against the real predators in the woods and survive to mouse our property. We could have never guessed they would outwit or outrun every hawk, fox, wolf and lynx out there and survive fourteen years, and counting, patrolling our property.

With the younger kids dressing up the kittens and the older kids fetching tools, we continued installing posts and beams and rafters for the rest of our house. Snow was a possibility in September and would definitely arrive by early November, so we only had a few months. We had to focus on the house structure. The inside of the house would have to wait and so, aside from building a couple two-by-four bed frames and some plywood countertops, our interior was just like the exterior. It was dirty and muddy and full of roots and stumps. It was more of a terrarium than a house.

One evening, the girls called me over to their room (which was just a different corner of the house with a privacy sheet).

"Daddy, something is glowing under the bed!"

I tried that dad thing, "No it's not, girls. Go to sleep!"

But they wouldn't relent.

"No, Daddy, there really is! It's glowing in the floor!"

We had zero lights in the house and there were zero lights outside of the house, and so it was pitch black. I used a headlamp to navigate over to their bunk beds.

I turned off the headlight and asked, "Ok, girls, where is the glowing thing?" waiting, unimpressed.

It took a few seconds but soon the girls burst out, "There it is, Dad! Do you see it? It's right there!"

And sure enough, as our eyes readjusted to the dark, a handful of abstract glowing shapes on the floor appeared under the bed. I flipped the headlamp back on and it all disappeared. There was nothing there. There was just dirt and sticks and chunks of old rotten log.

"What!?" I said, talking out loud to myself.

I turned the light out again and this time got down and put my hand on the largest glowing object.

"Turn your headlamp on, girls," I instructed, and as the light illuminated my hand, it revealed a soft wet piece of rotting aspen wood. I didn't know anything about bioluminescence but, here we were, in the pitch black of our new house, learning together.

Getting to work in town was a bit of an event that first summer. I would throw on a full set of rain gear and quad a mile through the bush, pinning it at the swamp crossing. I'd stash the quad near the truck in the abandoned field. In the truck, I'd throw it in four-wheel-drive and beeline it straight towards the impassable puddle. Once I was on the gravel road, it was an easy commute to town.

I had a quick shower at the office, washed off the mud, got changed, and no one knew. I was a professional technologist for eight hours, transforming right back into a muddy, mountain man by the time I got home. And so I made some money, Rose raised our kids, and we built our home in the evenings and on weekends.

It was hard work. It was productive. It was rewarding.

It was a great life.

The wet summer gave way to an even wetter autumn and the right-of-way that was to be our driveway was a disaster. With the stumps and topsoil scraped off, all that remained were high spots of sticky clay, sitting like islands amongst small lakes. Our access was not getting easier. The only way it could improve now was to get an early cold spell and freeze the ground solid.

When you're hoping for one particular weather pattern, it becomes evident just how many actual options there are. As evidenced through the travail of millions of farmers throughout time, the weather does whatever the weather wants.

In late October, it snowed two feet – two feet of fresh snow on over-saturated clay. It was a tiramisu of destruction. Deep snow on deep mud is the epitome of "high-likelihood" and "significant-consequence" on the risk matrix. It's the worst-case scenario. Nothing can drive through two feet of snow, and nothing can clear the snow because nothing can drive through two feet of mud.

Our construction plans were in rough shape. At this point, we had managed to finish framing the roof and walls, put a thin layer of spray foam on the outside, and clad it with cedar board and batten. We had installed an old wood cook stove, purchased via classified ads for one hundred dollars, and an old wood stove heater, another hundred-dollar second-hand find. After a long, damp, cool summer and the soaking-wet fall, the wood heat was heavenly. We were comfy but were still missing a few things: water, electricity, a floor, windows, interior walls, a bathroom, a kitchen . . .

We needed to make a major decision.

With winter looming, the house incomplete, and no driveway, how could we possibly stay here? Maybe we could just find a place to stay in town for the winter? It would be a nice break from our six-month off-grid adventure, but it would be expensive. It was very tempting, but we had been tempted by normal things before and, at this point, I think we had just outgrown it. Rose was uncertain how we would do it but she was very trusting. We would just find a way.

The ground slowly froze underneath the insulating snow. It seemed to take forever but, by mid-November, the rumblings of a giant piece of steel were heard coming down our driveway. An ancient, green Terex bulldozer was clearing the snow. Our road builder had failed to get a road built so far, but he did not leave us totally stranded. Agents of chaos – mud and snow – when subjected to freezing temperatures, fall right in line. With a hundred percent compliance, the snow rolled out of the way and the mud sheared to a smooth, flat surface. What was impenetrable yesterday, we could drive with a Honda Civic today. And not a moment too soon.

Harvesting firewood and packing it back to the house wearing snowshoes and pulling a sled had been slow work. Now that we could drive the truck to the house, we could get a whole load of firewood at once. We could also finally get the large pieces of glass for our south-facing windows and replace the poly plastic that was still stapled over the openings.

At some point in the fall, the muddy floor had become a bit much, so we repurposed the roof tarps and laid them down stapled tight to the walls. I had clearly never considered this stage of construction. We were in figure-it-out-as-you-go mode. Of course, the floor was still mud and sticks underneath and the ground was a foot and a half lower at the back door than it was at the front. I had already installed some premium plywood countertops for Rose at their finished height but, in order to reach them, we had blocks and planks to stand on, as the floor was low in that corner.

We dreamt about a level floor. With the driveway in shape, we could hire a truck and bring in a whole load of gravel. So we pulled out the tarps and pick-axed the top six inches of roots and clay and topsoil into wheelbarrows to be thrown up onto the living roof or heaped up against the house. With the virgin clay exposed, we started wheeling in load after load of gravel. A couple of small dump trucks and hundreds of wheelbarrows later, the gravel was approximately flat and level throughout the whole house. It was like a Christmas miracle. With the tarps laid back down and stapled to the walls, we even covered the living room floor with those interlocking yoga mats for a truly finished look.

The hovel took one step closer to being a house and it felt like we just couldn't stop winning. December twenty-third we finally received the large custom window glass and, for the first time, looked clearly through the front windows. It was a snowy wonderland and it was beautiful.

For power, I salvaged five old 12-volt batteries and connected a three-hundred-dollar RV solar kit – for a total of nowhere near the advertised sixty watts. The only electrical load we had in the house was a cell phone charger and a couple of one-watt LED puck lights. The tiny solar panels were just fine for now.

By this time, we were on our second sanitary system. In the summer, we had built an outdoor flushing toilet with an integrated septic tank and field. I dug a hole about the size of a forty-five-gallon drum and, on the backside, built a high-level outlet into a three-inch ABS pipe. The pipe, running just below the ground level, ran into a couple of infiltrators which were covered with soil. On top of the hole, which would become the septic tank, I built a wooden platform with two-by-six and plywood and installed a normal flushing toilet. The contents of the toilet emptied directly into the hole in the ground and the eventual overflow of water was intended to drain into the infiltrators.

A tiny problem was that we had no running water. But we did have a creek only fifty yards down the hill. Using the toilet required a run down the hill to fill up a water jug, then back up to the potty, fill the back of the toilet, and it was ready to flush as required. This toilet worked as designed but had a few shortcomings; there was no building to enclose it and it lacked functionality in freezing temperatures, which was most of the year.

So, with winter upon us, we employed a new-fangled indoor composting toilet kit ordered from some far corner of the internet. It was terrible in all respects and got worse with age. It was a forty-five-gallon drum with an industrial plastic liner. At the bottom of the drum was a grate that would supposedly hold all the solids on top and let the liquids through. On the top of the drum was a custom cover that included a toilet seat.

It had the appearance of simplicity – you could just go and then throw in a handful of sawdust. But the operational details got murkier than that juice at the bottom. Also installed on the cover was a hand-crank water pump. Below the grate, at the bottom of the barrel, was a pickup tube. The output simply returned the liquids to the top of the pile. It was kind of like a hand-operated chocolate milk fountain, except for poo water. The whole thing looked silly and, sitting three feet off the ground, felt weird, but otherwise it seemed to be working. It did smell a little funky and I installed a little, twelve-volt fan and plumbed it straight through the wall.

It did not work.

Within a couple weeks my concerns were piling up. The pile was adding up and, if I extrapolated it out, well, the math didn't compute. It was supposed to last us six months before having to empty it. The advertising literature was not exactly clear – it didn't specify its total capacity in any meaningful terms. Either way, the plain truth was it was not designed for full-time use by a family of seven, six of whom did not regularly leave the home. Maybe for a family of four who only use it on the weekends – maybe then it would have lasted six months.

But here we were, in the dead of winter with a sweet-smelling barrel of human waste stacking up in a hurry. Having considered the logistics, I concluded the girls were peeing too much and all the liquid in the barrel was inhibiting the solids from composting and breaking down. We had to get rid of the liquids. And so, instead of circulating the liquids, I MacGyvered the pump output to fill a separate five-gallon pail. I questioned my life choices as I filled that pail for the first time, and so did Rose.

"What are you going to do with that?" she asked.

"I'm going to throw it in the woods."

"But it's all snowy. You'll need snowshoes," she continued.

"I'll just throw it off the driveway. It'll be gone by spring."

In all my dirty jobs, I had never handled concentrated human waste. I carried that bucket out the front door, down the path towards the driveway and then a little ways down the driveway. I stood in front of the snowbank on the far side. This would be a good spot. I launched the contents of the bucket over the snowbank so as not to leave a big stain, and was relatively successful. After a few more buckets I was on a roll. This wasn't that bad, and surely now our central-planning-style toilet would compost and shrink in size.

It didn't.

A week later, with the barrel brimming, I brought home a shiny new heavy-duty dolly. After pumping out as much liquid as we could and carting pail after pail down the driveway, we heaved and wrenched that surprisingly heavy barrel onto the dolly. The axle bent and the wheels squished flat but we struggled that thing out the door and down the path. It felt like we were disposing of a dead body. We shovelled a new little trail into a little corner of the forest that we'd never been to before (and didn't plan on revisiting), wheeled the barrel, pushed it over and gagged at the violent eruption of odour.

"I can NOT do this!" Rose warned, as she reeled backwards.

"It'll just take a second," I said, grabbing the bottom of the barrel. "C'mon, I need your help!"

Rose scrunched up her face and tentatively stepped forward. We pulled and heaved the bottom of the barrel, jerking it to dislodge and empty the contents.

"Oh, gross!" she muttered and ran back to the house.

We had reached a magnificent new peak in humility, and yet, like the widow's jar of oil that overflowed, this toilet was not done giving. The winter is long at the homestead, with snow for at least six months and just three months being reliably green. In late winter we were still struggling to extend the service life of our indoor composting toilet.

I needed to empty that juice again. I had successfully done it a few times now; it was becoming routine. But today was different. The moment I launched its contents into the snow, my whole life changed. I didn't truly understand the snow in this region, and we had an unprecedented amount of it – over five feet. Being late winter, we had experienced the first thaw, and the undisturbed snow had a heavy crust on top while disturbed snow had turned rock solid. The snowbank in front of me was taller than the last time and was no longer fluffy and soft. Previously it melted away like cotton candy as I showered it with our homemade hot sauce, but this time was different.

Following through with the empty bucket, I froze in horror as five gallons of putrefied, brown, human waste moved in slow motion through the air, struck the peak of the snowbank and didn't make a dent. It was more like an ice-wall than a snowbank and, like a can of paint dropped on a concrete floor, the backsplash was magnificent. Effluent came raining down from above. Shock seized my body. I could do nothing but stare wide-eyed in disbelief as the wave of toilet juice crashed over my entire body.

My primitive response was to let out a death cry, "AAAAAAAHHHH-HHHhhhhhh!" which I immediately regretted.

The drips on my face flowed over my open mouth and my lips felt it at the same time as my nose smelt it. I instinctively stuck out my arms and legs like a starfish so no poopy part of my body would touch another poopy part of my body. My hands were not clean enough to wipe my face. There was no unsoiled surface on my shirt to wipe off my hands. I wasn't even sure that I could use my mouth to speak. But I had to do something, so, with foul drips dropping from my eyelashes, I lowered my face to the ground the way you would talk in the shower and yelled for Rose.

"Rose! Rose! Rose?!" I couldn't move and so I stood frozen, yelling, "RRROOOSSSSSE!"

Rose came running and the girls came running. About ten feet away they all stopped in their tracks. They had smelt it and, once they stopped, they could see it.

"OOOOOOOhhhh GROSS!" was all they could muster. Then, to make my suffering complete, they just laughed. And they laughed so hard they couldn't even speak coherently.

"Stay back girls," Rose finally sputtered. Turning to me, she asked, "What happened?"

"Rose, help me."

"I can't help you."

So I stood there for 10,000 hours, dripping, mouth-breathing so I wouldn't smell myself. What could Rose do anyways? We didn't have any running water. We had literally been melting snow for our domestic water use and that wasn't quick.

"Run back to the house girls," she demanded.

Then turning to me, suddenly all business, she ordered, "Take off your clothes!"

Thankful for the advice, I stripped down naked in the snow. I was still disgusted, with the smears all over my body, but grateful to be making progress. Rose ran to the house and fetched a box of baby wipes.

I used the whole box.

Even after an eventual wipe-down with real, warm, soapy water I still had that funk about me. It was the ambient old poopy-diaper fragrance. Not like an intense fresh poop smell, but a disconcerting sneaky smell that betrays contamination. If humility was the game, I had just won in a blowout.

We had survived our first four seasons at our off-grid homestead and it had been a spectacular combination of highs and lows. The lows could be categorized as uncertainty and discomfort, while the highs were the adventures and discoveries, new skills, rewarding and productive work and, ultimately, conquest.

It was the most challenging year we have ever had. We were prepared to address the discomforts and inconveniences and slowly design or build them into the past. The kids never complained for a moment. They were the true "canaries in the mine." We watched them to make sure we weren't pushing too hard. Little kids need the obvious stuff, like snacks and naps and hugs and kisses, but they're otherwise pretty low-maintenance. They don't care about carpet in the house, or a bunch of plastic toys, or running water for that matter. They don't care about having a washing machine or clothes dryer. They hardly care about wearing clothes at all till they're three or four years old. And our older girls – who were five, seven and nine – were mostly concerned about helping out, playing with kittens, reading books and exploring the woods. Kids are a pretty good indicator of what's important.

I was about to turn thirty-two, just three years and a few critical decisions away from Freedom 35.

Keziah was very skeptical of the forty-five gallon toilet

Toilets are a Big Freaking Deal

Rose and I both grew up in normal houses with flushing toilets, but we have both been perfectly happy without them for the last fourteen years. Except for the poop-juice-pail toilet, our foray into composting toilets has been a wild success. I'm not an anti-flusher, but there are a few gigantic problems with the flushing toilet.

First, they use an obscene amount of precious water. In a normal house, toilets use about a third of the water, which is significant, but doesn't tell the whole story. See, in a normal house, everyone wastes water all the time. You run the water while you brush your teeth. You run the water while you wash your hands. You run the water to let your shower or tub warm up. You run potable water on your lawn and to wash your car. With a bit of incentive, it would be easy for a regular household to cut its water use in half by simply being mindful. For example, you can adequately rinse your toothbrush in half a cup of water – that's about 125 millilitres. Likewise, you can wash your hands thoroughly with about one cup of water. What you can't do is flush the toilet with a cup of water. You can barely flush the toilet with a couple of gallons of water.

In places like Hudson's Hope, where households commonly haul water, people are used to conserving. A little sign in the bathroom might inform visitors, "If it's brown, flush it down; if it's yellow, let it mellow." But this strategy has its physical limits and there will still be multiple toilet flushes per person per day. Our family can easily make do with about two gallons per person per day but, with a flusher, we'd use at least ten times that.

The end result is that a flushing toilet requires a much more prolific water source.

It also requires a septic system, which is problem number two. Septic systems, if done legally, range from really expensive to impossibly expensive.

The deal with human waste is that it should be kept from contact with humans or animals until it is broken down with biological activity. This is the goal, or should be the goal, of every human waste-handling system, even in giant cities. The benefit of the flushing toilet is that it transports the waste easily from your house to somewhere else. The problem is that a flushing toilet turns a pound of readily compostable matter into fifty pounds of toxic slurry. The bacteria that would naturally break it down in the soil can't survive in water, making it harder to process.

Our current outdoor composting toilet is a spacious, thirty square feet of floor space. All the composting happens below the floor within this perimeter. There are no moving parts and nothing to break. Twice a year, we empty out material that has been broken down for at least six months. It looks and smells like a compost/mulch mix you might put in your garden. We don't have any particular need for the composted soil, so we just spread it on the forest floor right there, throw grass seed on it, and bid farewell. The only indication, after more than a decade of use, is brighter, taller grass. This could continue forever, benefiting the soil even in a small backyard.

Compare that to a typical rural septic lagoon. Lagoons are used in areas where the ground might be too rocky or nonporous and can't absorb water. The flushed contents head straight into a multi-chambered septic tank, and the overflow, similar to a city sewer system, drains into a large pit or lagoon. These lagoons can be in the 5,000 to 10,000 square-foot range and certainly have a larger footprint than the house that feeds them. A healthy lagoon doesn't smell, but it must be fenced off because it is a drowning hazard for children and pets. I've never tried it, but apparently the anaerobic decomposition creates a lot of dissolved gases, making lagoon juice less dense than water and more difficult to stay floating in.

Just for the sake of a flushing toilet that's prone to clogging and overflowing, a 10,000-square-foot hazardous waste of space is a big price to pay.

That brings up the third problem with flushing toilets – their failure rate is much greater than zero. I've been to third-world countries and I've seen exactly what happens when toilets don't work consistently. An engineering friend of mine once told me that the difference between the third world and the first is clean water and toilets. I imagined there were a million things separating the two but, over time, I've come to agree.

Access to clean water and functioning sanitary systems may be the giant gap between a successful and failed society. If you've ever experienced a flushing toilet malfunction, it was probably an unfortunate experience. Certainly, if your children are in the least bit scientific, they have hypothesized and then tested the flushable limits. Anyway, composting toilets don't fail like that. It's true they can slowly fill up, but they'll never flood your bathroom floor – except the one I told you about earlier. Please don't use that kind. And that highlights the only real problem with composting toilets. They're not well understood, they require research, and there are some flaky systems on the market.

For a while, we tried the toilet box with a five-gallon pail method and didn't really like it. It works great for occasional use, like a weekend at a cabin, but with a family of seven it was too much material handling for me. After a few years, we built a magnificent two-chamber outdoor composting toilet. The key is the two chambers. When the first one is full, it is shut down, and the second one is used. The contents of the first chamber are allowed to compost for months or even years until the second chamber is full. Only then, after it has had plenty of time to decompose, are the contents of the first chamber shovelled out. With this glorious technology, there is zero handling of fresh waste.

Instead of an enclosed little room, we built ours like an alcove with only three walls and an open side facing the forest. With this design, fresh air is abundant, and smell is a non-problem. I would absolutely recommend this design for anyone in a similar situation to ours and I would definitely do it the same way again.

There are a couple of downsides – notably mosquito season and winter. Reasonably self-explanatory, the solution is to keep things brief and avoid unnecessary suffering.

We eventually made a Youtube video on this exact topic because it was a very important fork in the road for us. On our property in Hudson's Hope we had taken the normal approach, the first steps toward a giant debt.

On this property, we chose to collect rainwater and use the bathroom outside. Both a well and a septic system could each individually cost more than our entire property. In addition to the sticker price, they would force us into debt easily, pushing the final cost into the quarter-million dollar range. In a town, these services are already included and there is no option. Only by developing our own off-grid homestead could we opt-out and take a different approach.

We had clean water and a toilet; it just wasn't running or flushing. And a quarter million dollars could be the difference between Freedom 35 or getting sucked back into the rat race.

I should also say a bit about grey water, commonly understood to mean wastewater that does not contain human waste. It is still not officially recognized by regulations and bylaws of many municipalities, but there's nothing inherently dangerous about grey water. It is easily broken down by microbes in the topsoil. In our case, the grey water came from the kitchen sink, the bathroom sink, and the shower. It was just a few hours' work to dig a trench six inches deep and forty feet long behind the house. The two-inch ABS drain pipe entered an infiltrator that we heaped with topsoil and mulch. That system has worked perfectly for over a decade.

Glorious Independence

That first winter at Tipper Creek we had a couple of our lowest lows, but also made one of our biggest and most influential decisions. We had managed to buy our Tipper Creek land without a mortgage and had been building frugally to avoid borrowing. But we still owned the Dream Property in Hudson's Hope and it had a mortgage. We had friends renting it and, while I thought the rent was fair, it did not cover all the costs of a rural off-grid property. The rental income didn't even cover the mortgage payment. The property still had a piece of our heart and so we were willing to put a couple of hundred dollars a month into it, but when additional costs came along – like having our well purged and a new pump installed – it was painful.

At the same time, we knew the Tipper Creek property would be consuming time and money for years before it got close to completion. My salary as a full-time employee at BC Hydro was just not cutting it. I had learned a lot and the compensation was good but something had to change. The financial aspect is maybe the easiest for me to explain but there was a much bigger reason. Working there made me want to stab myself in the face. I had moved around, seeking new challenges and learning new things, and I had worked alongside plenty of great people but, overall, the experience had become soul-sucking.

BC Hydro was like a beautiful woman who just didn't love me – all the benefits, none of the love. It was like dating a chatbot. Being a crown corporation was a big part of it. A change of government could suddenly throw out existing priorities and install new ones overnight. It was disheartening. It operated by mandate instead of surviving and excelling in the free market.

Beyond its underlying political nature, the union environment had its own challenges. Excellence was not recognized and failure often went uncorrected. The union agreement, a whole booklet, was so rigid it would not give any flexibility to managers or employees to optimize the workflow. A significant number of union workers viewed the contract as something to be gamed – it was always about what they could get. Most of the managers I had were fair but their hands seemed tied when it came to affecting positive change. The air was soured by a few workers who were straight-up dead weight.

I wanted to understand the big picture and make the whole business run better. I wanted to live and die by the sword – to be excellent and get promoted or fail and get fired – and I couldn't do that as a union employee.

I quit.

With a dirt-floored home and five little girls and my beautiful wife fully dependent on my income... I pulled the plug.

In the last year or two, retired technologists had begun to contract back to BC Hydro. Of course, they had no guarantee of work, but the rates they could charge were more than double. The retirees had the safety of a pension and they probably didn't live in a hovel, so for them it was an easy decision. For me, it was far riskier. But with risk came reward; if

I could keep busy we'd rake it in. It was a no-brainer. And so, for $325, I incorporated a company, designed a logo, paid $1,500 for commercial liability insurance and I bought a shiny new iPhone 4.

I was in business, or so I thought.

On my last day at work, BC Hydro threw a big wrench into the mix.

My immediate manager wasn't bothered by my transition from employee to contractor. He was a smart guy and the way he figured, without all the bureaucracy, he could hire me back for half the time and get more work completed. But a few other managers got wind of it and saw it differently. They thought if a ten-year veteran could quit and become a successful contractor, then many would follow and they could lose a lot of their workforce. The director of operations got involved and, at the behest of some of the most vocal frontline managers, gave verbal instructions that no one in BC Hydro could hire me back. His plan was to boycott me. I was told this verbally, but no one involved would put it in writing. Of course, it would be illegal to "blacklist" me but, with an unwritten rule, how could I prove it?

The tactic backfired and cemented my decision.

Unfortunately, determined or not, I wasn't about to make any money anytime soon. I started making phone calls and I had a few leads. 2012 was a pretty decent time to have marketable skills, but nothing was going to happen overnight. It was in this humbling circumstance that I made a call to an old friend back in Abbotsford.

Rose and I had met Wes and Jen when we were first married. We joined a small group together at the church that we affectionately referred to as the Good-Looking-Young-Married-Couples Club. Wes managed a construction team and I knew he could find a way to get me out of a jam. He assured me he had some work and I could start on Monday.

We packed up the girls and took a road trip down south, squeezing into my parents' place. It was a real throwback to work labour again. It was kind of refreshing to just work, pushing dirt around, no politics at all. The pay was a throwback as well, but it kept me busy and the money was better than any of my other options at the moment.

It took about three weeks before I got the call I had been hoping for. We got our first contract! I would provide construction management for AMEC, a large international EPCM (Engineering, Procurement, and Construction Management) service provider. I would be a hundred percent on-site, overseeing the construction of a brand-new, 138 kV substation for one of their main clients, BC Hydro. As an independent contractor, I would be boycotted, but with an international corporation larger than they were, apparently they wanted to abide by the law. They made no objection to my assignment.

It had been one year almost to the day that we had moved onto our raw land at Tipper Creek and now we were packed up and heading to the tiny town of Stewart, BC.

The decision to head out on my own changed everything.

We started earning some serious money. That first contract resulted in 2,000 hours of more-than-full-time work. It finally returned us to positive cash flow even while continuing to develop the homestead. It was also a whole new game for me to learn and I loved the challenge.

My scope was broad and included every aspect of construction – from earthworks, soil compaction, forming, rebar, and cement-sample testing through steel-structure erection, overhead high-voltage electrical bus work (conductors), steel-clad control building, right-of-way clearing, transmission line construction, safety, environment, operations, public consultation, client relations and contractual obligations, etc. Besides learning a million things, I was also free from the restrictive job description and union agreement. If something made sense, we did it. If it didn't make sense, we didn't do it. It was even more refreshing one day when I asked the whereabouts of one of the project engineers.

I was told, "He wasn't working out. He was let go."

I was pumped.

"We can fire people when they suck?"

This was terrific! This was the way. If they don't belong on the A-team, we fire them.

The move up to Stewart was a fabulous adventure for our family. I worked almost seventy hours a week, but I was only a few minutes away from our rented house so I could still see the girls each morning and evening. They settled into the rental in the middle of town and continued their homeschooling. It was a novelty to have power, water and a floor in the house, and be just a block or two away from a grocery store and a park. We were like normal people for a while.

Stewart is postcard-perfect. Snow-covered Coastal Mountains shoot straight up on both sides of a flat little river valley right at the end of a fjord-like inlet, the Portland Canal. The ocean is right on the doorstep. The little dead-end highway to get there passes by the Bear Glacier, which reaches almost to the road, and many actual bears, black and grizzly, are often seen mowing the spring grass on the road's shoulders.

We often walked down to the dock in the evening to throw in a prawn trap. The girls would check it the next morning, usually pulling up fifty to a hundred shrimp. It was a magical summer.

Sarah with a prawn on the Stewart docks

One of the brief trips home in early fall turned into a real heartbreaker. We were always excited to get back after weeks away and, even after a ten-hour drive, it felt great to jump into our own beds. But that night, getting in late with half-asleep and half-crying kids, we discovered the house to be co-opted. Rose went in first and, as I came through the front door carrying two little sleepers, she steamed towards me and spat out, "There are mouse turds! On the counter!"

There was nothing I could say and, while I stood silently, she looked around and continued, "There are mouse turds on our cups and plates and bowls and cutlery!"

"They're in the pantry! They're all over the floor!" she said, wheeling around, her voice rising.

It was enough to crush the heart of any exhausted mother but, if we could just crawl into bed and have a good sleep, we'd clean it up in the morning. I trudged into the kids' room and Rose stormed ahead of me, like an avenging angel finding an unfaithful lover in the act.

"There are mouse turds on the bed!!"

She flipped back the covers revealing little chewed holes and more turds.

"They're IN the beds!" she cried.

The tears flowed freely as Rose fully internalized the invasion. More than a little gross and unsanitary, a mouse home invasion is like a personal violation. And so, feeling betrayed and overtired, Rose began to rip off all the bedding. While she did that, I saw a little tyrant scurry across the floor.

"There he is!" I yelled and lunged at it.

He reached the edge of the wall and disappeared under the tarp and scurried across the room, a little bump moving under the tarp and then out of sight.

"Over here!" Rose called.

As I turned to look, I saw another. The more we looked, the more mice we saw crawling under the tarps in the living room and in the kitchen – it was a total failure. Physically and emotionally damaged, we set traps and crawled into stripped down beds. Instead of accepting blame, I placed it all on the tarp floors.

Who's idea was a tarp floor anyways?

Entering our second winter, the homestead benefited from our newly-improved financial situation. We swapped out our hundred-dollar cozy wood stove for a $3,000 Blaze King Princess. It used half the wood and only needed filling once or twice a day. Pure luxury. Dominated by the cold and damp summer and fall of the previous year, and the firewood-hungry, stone-cold-in-the-morning wood stove of last winter, the Blaze King won our heartfelt appreciation.

The driveway still suffered from poor drainage and regular flooding but winter brought a few months of smooth reliable ice road once again.

As soon as we could, we brought in loads of sand and gravel and a pallet of Type-10 Portland cement. Whenever we had a few days, we would clear a "room" (still just spaces separated by curtains) of all contents and prepare it for some civil work. We ripped out the tarps, installed electrical conduits and grey-water plumbing pipe as required, and graded and levelled the gravel. Next, we installed a solid layer of one-inch rigid foam and wire mesh propped up with little stones.

The girls helped with everything. They delivered tools. They raked the gravel. They tied the wire mesh together with the rebar tie wire and, most importantly, they helped me make the concrete. Already good cooks, thanks to Rose's mentorship, they had an eye for just the right slump.

"That's too much water!" they would tell me.

"You need more water, Dad," they'd instruct the next time.

We all shovelled gravel and we all wheelbarrowed it from the mixer on the driveway, down the path, through the front door, into the house. It took a few months but, one wheelbarrow at a time, we transformed our mousy hovel into a proper house. I see why people often start with the floor. And it's not that I set out to build a house that didn't have a floor. It's just what happened. Without a floor, you can't build walls, and without walls, you can't build things that are normally built into walls, like closets or shelves. Without a floor, you can't very easily build kitchen cabinets. Without a floor, everything that normally sits on the floor needs little bricks, boards or pucks to sit on. But with a floor – oh my, the things we could do.

Christina with a Pink Salmon from Hyder Alaska

Off-Grid Systems

The house was now the most deluxe tent we'd ever had. From the pop tent, to the tarp house, to the muddy, tarpy, mousy-floored hovel, we were now living in luxury. I upgraded from our RV solar panels to a proper 225W panel and added a charge controller and inverter. We had ourselves a power system. The living room and kitchen each had four one-watt LED puck lights and each room had one or two. We wired in 12V cigarette-lighter-style outlets and used USB adapters to charge our phones.

One of our most important pieces of electronics was a cell phone signal booster. Without it, we would only get a random bar of service but, with it we could usually hold a conversation. The inverter was turned on only when Rose wanted to use her kitchen mixer. If we needed to use power tools or top up the batteries in cloudy weather, we used a 3,000W gas generator. Even though the power system was tiny, it was totally sufficient for our needs with the exception of one critical thing – the fridge.

For the first few years, we did not have a fridge. To begin with, we had a regular camping-style cooler out of the back door and, for most of the year, that was totally sufficient. In September, nighttime temperatures average fridge temperature and are sometimes even freezing. Depending on the weather that day, we'd move the cooler in or out of the house. By October, we could fairly reliably freeze ice at night to be placed in the cooler. And from November to April, refrigeration and freezing were no problem at all. In the spring, the process would reverse and really only left us June, July, and August without any hope of natural cold food storage.

The solution to our problem was simply to do without. As it turns out, a lot of the stuff we would have put in a normal fridge was not missed that much. If you weeded out all the sauces and condiments and leftovers, the real important things we wanted refrigeration for were meat and dairy. Fortunately, our tiny new solar system would crank in the long summer days, and we figured we could try some electrical refrigeration. Of course, we wanted to do it better than normal, so instead of buying a regular fridge, we bought a stand-up freezer. The logic was a stand-up freezer has much more insulation and would be more efficient.

The freezer, of course, freezes, but we were hoping for a fridge. To solve this, I installed a special outlet that had a programmable temperature-controlled contactor in it. It had a six-foot lead with a temperature sensor on the end that I routed inside the fridge. The freezer essentially would always run, but the temperature sensor would cut it off at 2°C and allow it to come back on at 4°C. It operated very well and served us reliably for a few years until our next upgrade.

Although maybe not necessary for everyone's lifestyle, a deep freezer is pretty essential for a hunter-gatherer. There's a salmon-fishing season and a berry-picking season and a moose-hunting season and, if you're successful in those seasons, you need to preserve a lot of product for the rest of the year. Rose did a lot of canning and dehydrating but it was hard to compete with the simplicity and unprocessed benefits of deep-freezing. There was no way we could run a deep freezer at home yet and it would probably be a few years until we had a sufficiently large solar power system.

Thankfully, our friends Ben and Jamie, who lived in town, graciously allowed us to keep a deep freezer at their house. For those first few years, it was a great treat to make the trip to town and not just pick up meat from the freezer but also put in some laundry, use their shower, and sometimes even spend the night. It was a great big party with friends, lights, power and hot water. What else could we want?

Alongside our slowly-developing electrical system, we slow-walked our water systems. In the early days, we scooped water out of the creek in the summer and melted snow in the winter. The creek water was pretty brown and straight out of a beaver pond, so it was nutrient-rich but not exactly potable. That first summer, we relied on five-gallon bottles of drinking water from town. The snow, however, made beautiful soft clean water. It's impressive how a five-gallon pot packed full of snow melts down to a paltry one gallon.

We discovered the best system was to fill two five-gallon pots and put them on the wood stove at the same time. Once it was mostly melted, one pot could be poured into the second and then taken outside and packed full again. Once on the stove, the other pot full of water would be poured into the pot of snow and then it was used to get the next bucket of snow. It takes a fair bit of effort to obtain usable amounts of snow water. And no matter how clean the snow looked, it always had at least a few spruce needles and other debris. The water had a smoky spruce taste to it.

The big water breakthrough for us came with the building of our wood-shed three years after moving onto the property. Strategically positioned right beside the house and with over 200 square feet of metal roofing, it doubled as a rainwater source. As soon as the roof was up, we cobbled together a timber stand and slid on a two-hundred-gallon poly tank. A simple sixteen-foot length of gutter led to a little stub of a downspout, directing water through a kitchen sieve sitting in the tank opening.

After just a few spring showers, we were water lords. Up until now, bathing either involved a hike down to the river, a trip to town, or one of those tiny little camp shower bags. Not anymore. Like royalty, we could now use one of the big camp shower bags.

Not only that, but with the purchase of a countertop Berkey water filter, we could simply carry a pail from the water tank into the kitchen, fill up the Berkey, and then pour a purified glass of delicious drinking water out of the spigot. We would eventually install a larger rainwater system but, to this day, we are still packing water to the Berkey as our sole source of drinking water. Of course, the water tank on the back of the woodshed has to be drained at the first early frost, so it's really only reliable from May until sometime in September. We had to melt snow in the winter for a few more years.

To an unenlightened observer, it might have appeared we were suffering from our cobbled-together collection of limited water supplies. I agree that water is incredibly important for any house and especially for a homestead. I knew a well was the obvious solution. But I had owned three other properties by this point and, on every single one, had either dug or drilled a well. Even with a tremendous amount of money spent, none of those wells were perfectly reliable and none of them provided great-quality water.

The well we had to redo on our first property got low in a dry summer and couldn't support irrigating a garden when you needed it most. And being a shallow well, the water could not be trusted to drink unfiltered.

Our second property had a drilled well and produced a decent five to ten gallons per minute. But the water tasted like pennies, smelled like sulphur and left a hard white mineral stain on everything it contacted.

On top of that, it began to silt up after three to four years, requiring an expensive service call with the drill truck and a new pump.

Our third property had an existing shallow well which ran dry within weeks of us taking ownership. When I peered down the hole, I found a dead mouse floating on a few inches of water. We proceeded to drill over 1,000 combined feet in two separate shafts and were rewarded with one-third of a gallon per minute 600 feet down. It required a very large pump and some very fat, very expensive electrical cable. That whole adventure cost $45,000 and did not supply enough water to irrigate.

No, on this property, we would try something different – partly to save the money, but mostly just to prove the concept and benefits of an unconventional rainwater system. Almost anyone can collect rainwater and, although it must usually be filtered, it will always be free of hard minerals. For domestic use, rainwater has been the best-quality water we've had out of four rural properties. It's also been the cheapest and easiest by far.

The key to a good rainwater system is the first flush – a simple strainer, like the sieve on our woodshed water tank, keeps out the chunks but doesn't do anything to keep out smaller or dissolved contaminants. Between rainfalls, a roof gets covered with dust and debris like leaves and pine needles, and also yucky stuff like insects and bird poo. At the next rain, many of these contaminants are immediately dissolved in the water and join the rush down the gutter and into your rainwater system. The first-flush system is a little reservoir that catches the first five to ten gallons – by far the dirtiest water. Only once this reservoir is full does it divert the rest of the water into your storage tank. There are a few different styles of first-flush systems; the key is just to have one.

Even a modest rainfall can lead to a huge harvest in water. Unlike solar, which requires specialized panels that only collect twenty percent of the sun's energy, nearly one-hundred percent of rainfall can be collected on many different readily available roofing surfaces. In our summertime, we expect about sixty millimetres (a little over two inches) of rain in a month. On our woodshed roof, this adds up to about three hundred gallons or 1,200 litres per month. On a 1,200-square-foot roof, it would be almost 1,500 gallons (6,000 L).

Over the years, we've become very comfortable at using around 400 gallons of water a month for a family of seven. This includes drinking water, washing, dishes, shower, laundry, etc.

For us, the challenge is not a dry summer but actually a long cold winter. Without any rain or any melt, our rainwater storage slowly depletes through December, January, and February.

We've done a number of different things over the years to meet the need but the simple solution is just more storage. The bottom line is we don't actually need that much water for our house. We don't have luxuriously long hot showers or fill a soaker tub. We treat those as actual luxuries and enjoy them elsewhere.

Our emergency shelter was starting to look, feel and function like a real home. Rose's kitchen had shelves filled with glass jars of flour and sugar and herbs and spices. Her countertop, made of good-one-side, three-quarter-inch plywood, smooth and finished with mineral oil, gave her plenty of room to work. She finally had a sink that drained and a fridge that worked. But the centerpiece of her kitchen was the wood cookstove.

In our first off-grid place, we had a propane range and oven and it worked well enough. But we had visited with a lovely older couple back in Cherryville who still used a wood cookstove in their kitchen. I had never seen one in action and was thoroughly impressed. The lady enthusiastically told Rose about all the things it could do but also listed some things it couldn't.

"It can't make pies," she said.

Well, pies are pretty important in our house, and so it was an agonizing decision to go with a charming wood-powered cookstove but forfeit all future pies. The fact that there are still functional wood cookstoves a century after they disappeared from the modern home is a testament to their quality construction and design. A wood cookstove perfectly embodies the type of home we were trying to build and so, with a prayer for future pies, we installed a cookstove.

There was a little learning curve. They definitely do not operate like an electric or gas stove. The heat is more intense on one side than the other and care has to be given to not stoke the fire too quickly or let it die down too low. Of course, Rose figured this out in no time and consistently produced award-winning foods. Keeping a steady fire was quickly learned, and rotating and positioning bread in the oven would ensure even baking.

She soon discovered that there was no real barrier to making pies – in fact, she could make fantastic pies. She just had to figure out where to put it in the oven, not too close to the side and to turn it every so often to get an even golden-brown on top.

The stovetop also has a special way to it. With no particular elements, you can put your frying pan or pot or kettle anywhere you want. Instead of relying on a number on a dial, you have to pay attention to your dish and, if it needs more heat, you nudge it closer to the fire. The monolithic stovetop actually allows you to cover it with pots and pans when you are in the middle of something big.

One thing that's always on the stove is a big stainless five-gallon stock pot full of water. Instead of running hot water, we quickly got into the habit of keeping a pot on the stove. If it was ever too hot, it could be mixed with some cold. We would regularly put a splash of boiling water in a bowl, top it up with cold and a bit of soap and sit it in the sink to wash our hands before dinner.

After dinner, Rose would dump a couple of gallons of hot water in the sink, top it up with some cold till it was just right for doing dishes. It really only took a few more seconds than it would with running water.

The other interesting part of a wood cookstove is the warming shelf. An electric or gas stove would never give off enough consistent heat to make it worthwhile, I suppose, but with a woodstove, it has many uses. You can put your plates there before dinner to warm them up or put fresh buns there to keep them toasty. If you make a cup of tea but then have to run out the door for a quick chore, you can put your tea there. It'll still be hot when you come back. We do all that, and we also use it to dry out our eggshells so they can be crushed fine and added to the garden. We use it to dehydrate herbs and old bread before it gets moldy, to be made into breadcrumbs. The warming shelf is so helpful and friendly, it's heart-warming just to think about it.

Rose, wearing her apron and with that glowing smile, prepared delicious meals, fresh bread and hot cups of tea, filling the whole house with joy. Our little cabin was far from finished but seemed complete.

Our third year on the property, we were really gaining steam. We had made some wonderful new friends, Aaron and Elisabeth Schalin, and Aaron agreed to do some work for us with his excavator. The road was still a soupy mess a lot of the time and needed some ditches and culverts. We also needed to prep the ground for a shop. We had purchased some big chunks of scrap culvert for dirt cheap and Aaron was able to position them in the bank for a future root cellar.

We finally brought our Wood-Mizer LT15 home after leaving it stranded in Hudson's Hope for a couple of years. It was put to use dicing up half a dozen large spruce, cutting the juicy fibres into timbers like birthday cake cut into big, delicious squares.

There's something about a freshly-milled stack of massive timbers. Having never before seen the sun, the fresh wood fiber is a bright white and has a lovely sweet smell of sap. Dave did most of the work while we were away, piling up fifteen massive beams.

Julia and Christina on the stack of freshly milled beams for our shop

As soon as the area for the shop was levelled, we brought in a few loads of gravel and immediately formed up footings. It was a couple of long days of shovelling into the mixer and we had 200 linear feet of footings poured – about 400 cubic feet or fifteen cubic yards in total. We had previously bought a lift of twelve-foot 8x8 fir, which would be perfect for the posts, and our spruce timbers would be the beams.

At the same time, we were also building the woodshed. Aaron sold me a lift of 2x6s he'd been hiding over at his ranch that made great rafters. And all the live-edge, one-inch slabs we took off the timbers made up the straps for the roof and provided the gappy siding I wanted for the woodshed. With a bit of galvanized tin, the woodshed was done by the end of June.

Julia milling a log with the Wood-Mizer

Somehow, with all that work going on, I also thought it was important to build a tree stand at the back of the property. I had built a few tree stands for hunting before but I had something different in mind this time. This would be more of a permanent structure, built with twenty-five-foot treated poles and big enough that a couple of us could comfortably spend the night. We built a ten-foot-by-ten-foot platform twenty feet off the ground with two-foot high solid walls all the way around and the rest open to roof height.

I figured since we were renting a machine with an auger and picking up a load of treated poles, we might as well grab some extra and get a head start on a future horse shelter and barn. Like so many of our projects, those treated poles intended for the barn stood like monuments for a few years before finally being transformed into Sarah and Abby's house.

While we were away in July, Dave was harvesting more big spruce that would eventually become the log rafters for the shop. Dave and Shoshanna were busy working on their place, much like we were, but, thankfully, Dave was also willing to offer me his log-building expertise. When we returned in late July, he had twenty-two beautiful straight twenty-foot-long spruce logs. They had all been peeled and shaved flat on the top.

Dave, Aaron, all the girls and I worked to lift each rafter up and set them down on the beams. The girls chained a rafter about the middle and hooked it up, Aaron hoisted it up with the hoe, and Dave and I stood on the beams and guided them into position. Once settled, Dave would scribe it, we'd roll it over, and he'd cut a flat notch to fit snug on the beams. With that complete, we'd flip it back over and auger a half-inch hole down through the top of the log and two-thirds of the way through the beam. Then we drove a fifteen-inch length of rebar through them both. We added 6x6 angle bracing in the corners, giving it strength and the awesome look of a timber frame. The roof was finished with 2x6 tongue-and-groove and, directly on top of that, a triple layer of torch-on roofing.

After two years of having zero covered space outside of the house, the shop was a game changer. The value of outbuildings had become painfully obvious over the last two winters with everything entombed in five feet of snow.

The long days of early summer led to significant progress and, with our extra money and time depleted, we shifted gears. The beginning of August is peak summer, complete with heat waves, river days, and late nights around the campfire. But, by late August, everything had changed. A few leaves already turned yellow and our first frost was a pending threat. The days were shorter and cooler and it was time to think seriously about winter preparation. I always plan for debilitating rains in the last half of October and permanent snow by early November. In a perfect world, we'd have all outdoor projects wrapped up and the woodshed full by mid-fall. September would be the perfect time to finish things up but, in reality, all outstanding projects are unceremoniously abandoned the instant hunting season starts.

Rose shovelling gravel into the cement mixer

Bonked Him Over and Chainsawed His Head Off

Hunting is so much more than just trying to get meat. Hunting is time to spend with friends and family. It's the impetus for exploration and wilderness adventure. It's the opportunity for me to go out with just one of the girls and teach her something about the woods. It's the way we take notice of the weather and the season and the time of day. It's how we find berry patches, precious little clusters of chanterelle mushrooms, standing-dead firewood and good milling trees for future projects.

Hunting is how we found the year-round spring that would eventually provide us water, even in the dead of winter. And it's how we found our fabulous private beach and the clay cliffs down at the river. The nearest edge of our property was only a couple of hundred yards from the river and it took just ten minutes to hike there. But that section required a very steep little trail and the river was mostly shallow without any usable shoreline.

During hunting season that fall, we found a bend in the river with twenty acres of gravel and a sandy beach. And the downstream bend had a curious outcropping of pure clay that resisted river erosion and formed a large pool with a circulating eddy current. It would be a fantastic fishing and swimming hole, and the clay itself was pure enough to create fired pottery.

Hunting was all of those things, and this year I had been drawn for an any-bull moose tag. The last two moose seasons, we'd had no special draw and, although we still got out, we couldn't find the mythical spike-fork. Both of those years we'd survived on a deer or two we harvested in November back up in Hudson's Hope.

This year, our hope was to find a moose right from home. A few years earlier, Sarah, who had just turned ten and got her first hunting license, had a close encounter with a bull moose within earshot of the house. We called it in together – to about twenty yards – and although he was the ideal meat bull, his antler configuration disqualified him from legal harvest. This year, Sarah and I both had moose tags, and we would head out for an hour or two each morning and evening. We could go in any direction – down towards the river, back to the abandoned pastures by the impassable puddle, up along the beaver ponds or just north into the rolling timber. It was late September when I figured the bulls should be getting responsive to cow calls. My cousin and his family – Andrew, Colleen, and the kids – were coming for a visit and would be arriving anytime. But there was just an hour left of dusk and we didn't want to waste it.

Sarah and I snuck across the creek and then up onto a little ridge spotted with large spruce and pine.

I let out a cow call, "Hhhhmmmmmuuuuuuhhhhhhhh," and heard an immediate response.

"UUmmmhhh, UUmmmhhh."

Sarah was wide-eyed. If we could get it into range for her, this would be her moose. We called again and, again, the bull grunted back. Soft at first, it was hard to tell in which direction he was and if he was coming our way or just staying put. We were trying to be patient and not stomp around looking for him but we only had another twenty minutes of light. Just then, we heard the rattle of gravel kicking up on a fender and realized it must be Andrew and Colleen coming in along the driveway. Normally we'd run out to greet them but that would have to wait.

We had to quickly get closer to this bull and so, making a little more noise than we wanted, we pushed through the brush off the side of the ridge to get a peek into the flat below. The bull was doing the exact opposite, coming up the slope to find us, and we bumped into each other in the alders. He was a big bull but we only saw him for a second before he ran back down the hill. We ran back up the hill to find a vantage point. I could see him crashing through the willows below but couldn't get a shot.

He was clearly desperate for love because, when I gave one last cow call, he stopped and turned. All I could see was his neck and head, the rest concealed by brush. And Sarah, not tall enough, couldn't see him at all. I would have to take the shot and I would have to take it fast. With one 165-grain Barnes solid copper bullet out of the 7mm Rem Mag, he dropped on the spot.

Sarah and I exchanged a "No Wayy!!" and rushed to locate him and make sure he was down for good before dark. Having never explored that particular area, we realized for the first time what a tangle it was.

From the ridge, it had seemed more meadow-like, dotted with great big birch trees, but now that we were in it, the meadow was mostly twenty-foot-tall clusters of alders and willows packed in between with chest-high bearberries. The bull was dead with one shot right to the neck. Now, fully comprehending the amount of work we'd created, we urgently headed in the direction of home. We anticipated a helpful circus atmosphere with Andrew, Colleen, Rose, and all the kids working together like a colony of ants to transport the beast back home. But that's not what happened. Our straightforward recovery plan would be dashed by an emergency rescue operation.

Rose and I had always dreamt of having horses on our little ranch. Rose had grown up occasionally riding horses with her friends and, as a teen, I had worked mucking stables in exchange for horseback riding. My interest most definitely arose from a desire to hunt on horseback. I had been to the backcountry on foot and loved it but hunting big game that far in the mountains required horses.

In Hudson's Hope, we came close to buying our first horse but, without any proper fencing or infrastructure, we wisely waited. As a testament and commitment to our plans, we did buy our first saddle.

During our two years in Cherryville, we came across a local hobby farmer lady who told us, "The best thing my dad ever did was buy me a horse when I was fifteen. It kept me out of trouble for years."

With our wilderness retreat taking shape and with Sarah now turning twelve, we felt we could finally make it happen. The girls had been taking riding lessons for the last year and they were all in favour. With some help from our horsey friends, we found a nine-year-old Canadian gelding for sale for $3,500. Nokee had been well-trained in western and, unconventionally, in jumping as well. He was well-mannered and seemed gentle with the kids. The owner, interviewing us as potential horse parents, warned us that he was a little rusty without much recent riding. She suggested a professional spend a few weeks to reaffirm his training.

"He likes to go fast. Actually, he loves to go fast," she said.

Fast is not exactly what we wanted for the kids but, all things considered, Nokee seemed like a good horse. Unusual for a Canadian, he was a chestnut brown and had a very beautiful long mane with natural blond streaks. The girls, Rose included, were in love.

I didn't realize it at the time, but the Canadian breed of horses has a long and storied history. As Canada was founded with competing efforts by the British and French, the growing population of horses was a combination of the best cavalry from both French and English stables. From the 1700s to the mid-1800s, that breeding stock led to a herd of exceptionally tough, easy keepers with large feet and big bones, all naturally selected by the brutal and unforgiving Canadian wilderness. They were not too large but very stocky. They were smart calm survivors that came to be known as the "little iron horse." Versatile as a saddle horse and a beast of burden pulling a plow or cart, they could do all things required on the frontier.

Their renown in North America led to almost complete decimation during the American Civil War. Like so many implements of war, good horses were in high demand and nearly all Canadian horses found themselves with a one-way ticket to the American South. Apparently, some of these horses eventually contributed to the Morgan breed but, either way, the distinct Canadian breed disappeared in the U.S. In Canada, their numbers remained low until a resurgence in the late 1900s, as their value as an all-around horse regained popularity.

The girls thought he was pretty and Rose and I thought he was safe. He was one of us, a great testament to the toughness of Canada. Besides, we already had a saddle; it only made sense that we finally get a horse.

We took Nokee to a professional trainer for a few weeks for a tune-up and scrambled to fence in three acres of pasture. In late June, we brought Nokee home and created a whole new facet to our lifestyle.

The girls loved Nokee and spent hours brushing him and braiding his hair, leading him on halter and, of course, riding. It became obvious fairly quickly that one horse is an awkward number of horses to have. As soon as the girls were comfortable enough to ride around, it was apparent they needed a riding partner. Also, Nokee had a great big pasture all to himself and, as a herd animal, he was not comfortable on his own. We looked for another horse, but they're expensive and we didn't yet have our own horse trailer. It was a lot all at once.

That was the state of things in mid-September when a windstorm dropped trees across our fence. We were gone at the time and Dave phoned to tell us that Nokee had escaped and was missing. We came home the next day and began the search. We figured he would be looking for other horses but, with none anywhere nearby, I think what really happened was he just ran scared. We could see his tracks down the driveway but once it hit hard-packed gravel road we lost his trail. We hiked and quadded and drove all the trails within miles of the house and found no sign. We put the word out amongst friends in case anyone saw a rogue horse, but heard nothing back.

It had been six days since he went missing, and we were devastated with the loss of our newest family member. We even called the country music station and asked that they put out a missing-horse announcement on the radio.

It was in the midst of this tragedy that Sarah and I went out and shot the moose. And at the very moment that Andrew and Colleen showed up after a ten-hour drive, Rose got the call. It was the radio station host. She'd received a phone call from a guy in the woods who claimed to have our horse.

Instead of welcoming our weary travelers with a big hot delicious meal, she just waved apologetically and pointed to the phone as it rang some strange guy in the woods. The guy answered, clearly pushing the limits of cell service and, in a choppy, brief, and perplexing conversation explained that he was standing in a clearcut with our horse. He was out moose hunting and was nowhere near our property. Rose was having a hard time understanding how it was possible he could be seeing her horse so far away but, at the same time, she was ecstatic to hear that he was alive.

That's when Sarah and I returned to find an exhausted Andrew and Colleen and a dumbfounded Rose.

Our original plan was to simply welcome our guests, have dinner and visit over a cup of tea. But now we had two other urgent requirements: we had to extract a dead moose before it spoiled and rescue a live horse before it disappeared once again.

"Ok, I'll need to deal with the moose and the big girls can help me," I reasoned. "Rose, you'll need to get Nokee and ride him back," I continued.

"I don't know if I can ride that far, or if Nokee can. . . and I would need someone to drive me out there," Rose countered.

Andrew chimed in.

"I can drive you, if that helps."

We all looked at Coleen.

"I'll stay here with all the little kids and work on dinner," she offered.

And so, in the dark, around 9:00 PM, armed with chainsaws, quads, and knives, Sarah, Abby and I began cutting a trail to the moose. It was slow going but if we cut a more-or-less straight line; it was only about 300 yards off the driveway and less than 500 yards from the house. We cut only the worst of the alder and willow and deadfall, and bucked and reared our quads over everything else.

Finally making it to the moose, the girls used all their strength and body weight to pry the legs apart and help me shift the giant carcass so I could gut it. By midnight, we started loading quarters onto the quad.

In the meantime, Rose and Andrew had mercifully found Nokee, but their original plan totally derailed. It had taken them over an hour to get there – a total of forty-two kilometers on forestry roads. Even if he was healthy, that was an impossible distance to ride, given the circumstances. Unfortunately, he was not healthy, with a noticeable limp. It seemed he had something jammed in the bottom of his hoof.

Rose called a friend with a horse trailer who graciously agreed to come rescue them. Rose headed out to the main road to meet her and Andrew, not previously known for his horsemanship skills, waited with Nokee.

It took hours and, thankfully, Andrew, at home in the woods, built a little fire, put his belt around Nokee's neck and hung around all night like a real wild-west cowboy. Poor Colleen didn't get any adventure and had to wait till the early morning hours before any of us returned. By 3:00 AM, the moose quarters were hanging in our new woodshed and Nokee was tied up securely outside the house. The lasagna was fantastic, capping off an equally fantastic rescue/recovery campaign. The fascinating irony was that Nokee's job was to pack out the meat but, instead, he hindered the hunt and required that we pack him out.

The next morning, we were all able to head back to the kill site to recover the head and antlers. It was much closer to the way I had imagined it – a happy crowd of people swarming the woods with the kids inquisitively exploring what remained of the carcass. Andrew and Colleen's oldest, Maya and Elise, the twins, were just five at the time.

They would later recount the story while at a friend's for a dinner party, stating, "The big man bonked a moose over and chainsawed his head off."

Sarah with our first moose from Tipper Creek

The Great Flood

I was thirty-three, and it seemed like we were on the cusp of our Freedom 35 plan. At that moment, it seemed our lifestyle was amazingly similar to the fantastic one I had described to Rose fifteen years earlier. We had a legitimate cabin and, after harvesting a trophy-class bull moose – enough to feed our family for a year – right from our property, it cemented our wilderness credentials. The girls were growing up with the glamour of wildflowers, kittens, and a handsome horse complemented by the dirty skills of cutting trails, gutting, skinning, and packing their own meat. We were on the right track and well past the point of no return.

For years, it seemed significant efforts were yielding small results but our progress this summer combined with the successful hunt clinched a major win for us. It seemed now that even modest efforts were returning huge results. Friends and family, who had politely kept their doubts mostly to themselves, noticed our transformed homestead with pleasant surprise. We had lots of development ahead of us and still had some outstanding debt, but we were close enough that success seemed certain. But the wild trail of success is never a straight line . . .

Through 2013 and into the early winter of 2014, most of my contract work was through AMEC. I estimated, performed design reviews, developed staging and commissioning plans, and performed testing and commissioning on site. Rose and the girls always joined me and we would all cram into little hotel rooms for weeks or even months at a time. Rose quickly adapted to tiny hotel kitchens, keeping us fed and continuing to homeschool the girls.

In addition, I took a few jobs for other clients and, on occasion, subcontracted others to help me. One interesting project was a new data center in Kamloops, BC. I had contracted a friend of mine, another ex-BC Hydro employee, and our two families spent a month together at the hotel. Their girls and ours spent hours swimming in the hotel pool and we often all went out to dinner together. In a lot of ways, it felt like a family vacation.

The work was interesting, too. Data centers were kind of a new thing and their electrical architecture employed novel solutions to ensure maximum reliability. Complicated protection schemes helped avoid over-tripping, and massive flywheels provided spinning reserve and continuous uninterruptible power supply in case of total grid failure. The project paid well and was a positive experience.

I was hoping to take on more projects on my own and, in March of 2014, I got a massive opportunity. I was asked to supply a team and lead the P&C testing and commissioning of the second largest independently-owned hydroelectric project on the BC Hydro system. The Forrest Kerr Hydroelectric Project, owned by AltaGas, was building three separate facilities all clustered in the remote northwestern coastal mountains.

To say the project was complex is an understatement; it was audacious in every aspect. It was hundreds of kilometres from the nearest BC Hydro power line, 1,600 kilometres from Vancouver, and fifty kilometres through rugged mountains to the nearest paved road. The main generating station captured water from the Forrest Kerr River just above its confluence with the Iskut River. An intake tunnel needed to be blasted through two kilometers of canyon wall to the underground powerhouse.

Subject to huge swings in water flow and flash floods during winter storms, the water was highly abrasive, carrying tiny suspended rock particles from its glacier runoff.

The underground powerhouse needed to be blasted out of the bedrock and hold ten 22.5 megawatt generators with Francis turbines. The second station, Volcano, was fed from the high-elevation Volcano Creek and required a three-kilometre-long penstock with one thousand feet of head feeding two separate ten megawatt Pelton turbines. The third station again required a tunnel, this time right through the heart of a mountain, to capture McClymont Creek before it tumbled down a series of cascades into the Iskut.

Each one of these projects required extensive and creative headworks to tame and redirect the water. An Obermeyer dam (a ramped steel plate with a rubber inflatable bag underneath it) was installed in the narrow canyon mouth on the Forrest Kerr. When deflated, the steel plate lay flat and the water flowed like normal but, when inflated, the plate angled up and raised the water level by fifteen feet. This created a small head pond that flooded a large settling pond built adjacent to the river. The size of a hockey rink but twenty-five feet deep, its purpose was to slow the water down and allow the particulate to settle out.

At the far end, the water entered the long rock tunnel to the powerhouse. A hydraulically-controlled spill gate ensured the settling pond and intake gates would not be overcome by floodwater. When I first visited the intake and witnessed the tens of millions of dollars of development there, I couldn't have guessed I would stand there just months later and witness its complete destruction.

Besides the civil infrastructure, there was a significant amount of electrical infrastructure required. BC Hydro would be building the $750 million Northwest Transmission Line from Terrace to Bob Quinn Lake. From there, AltaGas would build another transmission line to the Forrest Kerr substation. Additional lines and substations would tie in the

Volcano Creek and McClymont power plants. Our job was to direct the installation of equipment and wiring in the control rooms at each substation and generating station, test every connection and program and test each protection and control device.

The project had an initial budget of $1.2 billion and, by the time we got there, they were tracking to be over budget and behind schedule.

Bill from Boston, the project manager, was always ready for a fight. With over one thousand people on site and a daily budget burn of a couple of million dollars, we arrived at the frantic peak of activity. With six months' worth of work to do in the next eight weeks, I tried to gently explain the impossibility of the prescribed timeline. Our team represented the last stop before anything could be energized.

"We will meet the schedule," was Bill's response. It was less aggressive than his usual tone, but no less demanding.

"Bill, we're doing everything humanly possible but it's simply not possible," I informed him as the weeks rolled on.

"I don't want to hear it," he always said. "We will make it."

He regularly raked crew leads through the coals during coordination meetings, demanding solutions and crushing weakness. Grown men cried. Even though I continually brought him bad news, he never gave me any direct heat. We worked a ridiculous schedule to come as close to the timeline as possible – sixteen hours a day, seven days a week.

As we got down to the final week, our threadbare attempt to reach energization unraveled from all sides. The engineers had accidentally undersized the control cables in the design and the electrical contractor was digging up the entire substation to replace them. We stood speechless as they gutted the yard and rolled back days of recent progress. On top of that, it was now evident that not all the high voltage electrical apparatus would be installed in time, let alone tested.

With a lot of money, reputation, and ego on the line, Bill came to the control room and laid it out.

"Listen," he told me with an unusually soft tone. "The energization date is a contractual issue with BC Hydro and, if we delay, we're subject to massive financial penalties."

He led me out of the control room and away from other workers and continued.

"What do we have to do to get this station energized?"

I surveyed the substation and shook my head.

"It's not complete; we can't energize. The control cables, the missing circuit breakers, the untested transformers ... We're still weeks away from being ready."

We both stared at the substation; with no electricity it was just a $50M jungle gym.

"We're going to energize. One way or another we are going to energize this line next week, and..."

"Wait! Did you say line? Do you mean the station or the line?" I asked, pointing up to the incoming transmission line. Bill didn't answer fast enough and I continued "You've always said the substation, that we had to energize the station, but now you just said the line..."

He searched the sky for an answer. He was not really an electrical guy.

"The line."

"The line." I repeated slowly.

"The line." Bill echoed.

"But you've always said the station, we've been trying to get the whole station ready...."

Unbelievable! Did we just work 49 days straight to ... Wow!

Concealing my disbelief, I pointed to the line disconnect and suggested, "We can just leave this disconnect switch open and energize the line. The whole station will remain de-energized and we can continue testing. We could energize the actual station when we're ready, maybe in a few more weeks."

"Perfect," he said with a nod. "Thanks," and he turned back to the office.

I'd been living a lie this whole time. So much stress, all for a misunderstanding between Bill and himself.

Even after overcoming that first critical deadline, the project continued well into winter and on into the next year.

To access the underground powerhouse required a long walk down miles of dark tunnel or waiting for a ride in one of the approved access vehicles. Many of the local Tahltan First Nations had labour jobs on the project and one of them, named Gord, was a driver. As our team's hours were much more extended than the regular shift workers who took a bus down and back from the powerhouse, we often relied on Gord. He was friendly, so we often got to chatting on our trips to and from the underground in his spark-suppressed Land Cruiser.

On one particular trip late in the fall, he asked, "Have you guys seen the flood?"

"No," we replied. "A flood?"

"Yeah, the intakes are flooded. We can go there if you want," he offered.

"Yes! Let's do it," I insisted.

We were aware of the rainy deluge turning the camp parking lot into a small lake, but I was totally unprepared for what we found upstream.

We rolled over the last hill to a vantage point where we could look down over the headworks and, after struggling to register what my eyes were seeing, I blurted out, "No way! That's crazy!"

It resembled nothing of the site I had visited a few weeks earlier. I started to piece together the different infrastructure. I could see only the tops of structures and, even though the spill gate appeared to be totally up, water covered everything. I looked for the Obermeyer and saw nothing but a raging torrent in the middle of the river. I presumed it was down but, regardless, the water had overcome the settling pond walls and was ten feet above the concrete embankments. With the dam down and the spillway open, it should've been a raging river totally separate from the settling pond and the intake structure. Yet at this point, they were indistinguishable.

There was a high concrete walkway above the intake screens, the only separation between head pond and settling pond. It was still out of the water, and we walked along toward the spillway gate. From here, Gord and I could see the full extent of the destruction.

"Look over here," I pointed downstream to a twisted metal wreckage. "Isn't that part of the sheet piles?"

Huge amounts of earthwork and concrete installations had been undercut and swept away. There had been a fish ladder, a series of small-stepped waterfalls with little pools that would theoretically allow spawning fish to swim upstream. They would just have to make a series of one-foot leaps in order to circumnavigate the Obermeyer Dam. Although I don't think the fishway was ever fully in service, it was clear it would never again be in service. Many of the 4,000-pound concrete Lego blocks that formed the channel were in a jumble.

This was such a remote area of the province. No one ever lived there full-time or even worked there on any type of regular basis. There was no significant amount of logging or mining that occurred and travel was mostly restricted to helicopter access for prospectors and, perhaps, guide outfitters. There was no solid local knowledge specific to that part of the river or to that exact canyon.

BC Hydro had studied the Iskut River decades earlier and had provided hydrological data to the independent power producer. But standing there, looking at the destroyed infrastructure, it was clearly insufficient data. What we didn't know then was that this wasn't a particularly abnormal flood. A week later, before the damage could even be fully assessed, another flood of equal magnitude would hit.

Standing there in the midst of the downpour and watching millions of dollars wash downstream, Gord wryly suggested, "They should've just looked at the tree line."

My eyes scanned down the canyon and then up the valley and noticed that the water was just lapping at the roots of long-established timber. The current wasn't undercutting any mature trees; in fact, the only greenery submerged in water was low brush. We all looked at each other and slowly broke into smiles. The engineers on a one-billion-dollar project had just totally failed, and yet Gord was totally unsurprised. Just a bit of insight from someone who lived a little closer to the land could've saved them millions.

I already liked him but, at that moment, Gord earned a special place in my heart.

The project was a huge challenge, technically and logistically, stretching our physical and psychological limits. There were periods of intense stress and sleepless nights. The team worked extraordinarily well and, overall, we knocked it out of the park. Besides proving my technical and managerial capabilities, it led to larger expectations for my business. Our invoices were far larger and our monthly expenses were significant. Without context, the expenses looked ominous. We shifted from an independent contractor to a small business. I had worked way more than full-time, but so had Rose as our administrator. She handled the books and made all the arrangements and, together, we were making good money.

For two years, I had been driving our 2002 Ford Excursion or '94 Jeep Cherokee for work and so, for the first and probably last time, we bought a brand-new truck: a 2013 Ford F-350 crew cab long box. It was a sign of the times and a reflection of our work and success. It wasn't our primary goal to build a business, but here we were and more opportunities lay before us.

True Wealth

In high school, I imagined becoming an inventor and entrepreneur, making millions. I considered going to art school or becoming a preacher. I enjoyed math and science in the same way that most people enjoy things they excel in. When I was fifteen, I bought a book called "The Ghost in the Atom" while on summer holidays with my parents on the Oregon coast. I was genuinely intellectually curious about particle physics and that became my stated goal when I first enrolled in a science degree program.

I loved playing sports. I loved climbing mountains. I loved hunting and fishing.

And I wanted to do it all.

All those desires stood in line behind Rose. As a nineteen-year-old with a wife, I felt an urgency to take all my interests and opportunities and put them together in some practical fashion. Looking back, it's easy to see a lot of my early ideas still form part of my life. My love for the artistic finds expression through our *Gridlessness* YouTube channel. Science, technology, and entrepreneurship have taken the role of income-earning.

Even in elementary school, I found creative ways to make money.

Mr. Watkins, my fifth grade teacher, began each school day by writing a line on the blackboard that any naughty children would be forced to rewrite over and over as punishment for bad behaviour. The line was something like, "I will be a good listener and not disrupt the class when the teacher is talking." I was a common recipient of the writing-lines award, and so I preemptively wrote a sheet of the prescribed lines throughout the day. If I ended up on the naughty list, I just handed in my page of lines right after the bell and walked out, having paid my dues. But if I didn't end up with the prescribed punishment, I sold my page of lines to other bad boys. In 1990, while Vanilla Ice was big, I was selling pages of lines for two or three dollars each.

My brain offered me the same biochemical reward for making money as it did if I scored a goal playing soccer or hockey. In high school, I bought a pop machine, which earned me about $8,000 in net profit over a few years.

I was always thinking of fantastic and epic business ideas, which always failed to materialize. Yet, suddenly, my boring field services company was on a tear. Despite the success, it was becoming apparent my business philosophy was not typical. In the last two years, all my clients, partners and competitors just wanted more money. While it was a tempting goal, I had long ago rejected it. My naturally counter-culture inclinations urged me to decrease my need for money.

I observed it was impossible to have "enough money." If you have a ton of money, you will want two tons. We all know this, but most do nothing to correct it. Some do a pretty good job of finding a reasonable equilibrium, but very few actually buck it.

Let me be totally clear: money is not the enemy. I don't believe having loads of money is wrong or bad. I have witnessed some with incredible wealth stewarding it wisely to make the world better.

The problem is the deception and seduction of money that lures many of us away from the truly important things in life. At a fundamental level, humans need food and shelter, purpose and productivity, and a relationship with God and the people we love. Everything we need or want falls within these three categories. The first one takes money, the second one makes money and the third one is free.

The optimal life – for the individual and for humanity – is to strike the balance between all three. We should work enough to provide for our needs and allow us time to build and maintain our relationships.

The common definition of wealth only measures one category: the amount of money we have. It gives no weight to the quantity or quality of our relationships or whether we are living with purpose.

For example, while rich by the common definition and without a need to work, a trust fund child might struggle to find purpose and productivity. And all the riches in the world are a likely detriment in the pursuit of true friendship.

Alternatively, a hippy approach might focus primarily on friends and relationships (free love) but demonize capitalism with its productivity and material necessities. The result may be a lack of purpose and detachment from personal/societal responsibilities.

The workaholic's long hours take care of financial and productivity needs often at the expense of healthy relationships.

Our aim should be to have abundance in all three, such that we can share with others. With an abundance of money, we can give good gifts. If we are especially productive, we can train and teach others. And if we invest more of our time, we can be a better friend, father, son, or brother.

All that is to say, a critical pillar of our Freedom 35 plan was to make the right amount of money. We wanted to fill the fuel tank, without risking a freak gasoline fight accident.

Our goal was to live like millionaires without being millionaires.

I wanted plenty of food and shelter and material things, and I wanted to be useful and productive – an important part of a team and maybe even an outstanding contributor to society. And I wanted a magical relationship with my wife, kids who respected me, and a pile of friends to enjoy the whole adventure with. We had come so far and achieved so much but I didn't just want to get close. I wanted it all.

We were already hunting, gathering and gardening so much of our own delicious food and, with Rose's devotion as a cook, we ate like kings. Fresh, sweet, crunchy carrots out of our garden, a few snap peas and cherry tomatoes were tastier, healthier and more satisfying than anything we could buy at a downtown organic food fair. A lean organic moose steak cooked on a campfire was not even an option at a fancy steakhouse. Fresh buns and a homemade meal prepared and eaten with family is something money can't buy.

We had a super-cute cabin in the woods that brought me a smile every time I walked up the little path and through the front door. Our property was like an estate, meticulously appointed with mature trees, wildflowers and huckleberries.

Our family spent time together – exploring, adventuring, and generally developing shared interests. We had a terrific circle of friends – some just through the trees, like Dave and Shoshanna – and many others around the province. We had friends to hunt with and friends to fish with and old friends to catch up with.

And I was certainly feeling productive at work, more than ever before. It had been quite a ride and was a welcome change of pace. But life is dynamic and our trajectory could not remain static. It was not quite the right balance for the moment, especially given the amount of time I was working away from home. In fact, there were many days that I would earn $1,000 and then eat dinner by myself, wishing I could pay $1,000 to spend the evening with my family.

I knew to achieve the millionaire balance I was looking for, I would need to work away from home a little less. I was free to work as much or as little as I wanted, given our relatively modest living costs, except for one small contract I'd signed, enslaving me for the next thirty years. The monthly mortgage payments on the Hudson's Hope property were only about $1,800, but the commitment weighed on my free spirit. It took up space in my brain and distracted me from my passions, threatening my liberty.

It seems silly now, looking back, that we hung onto the marshmallow house for so long. At first, we didn't know if we'd move back there one day, and then we thought maybe we could just hang onto it until it paid for itself. But, at some point, we were just clutching onto it like Winnie the Pooh with his head in the honey pot. We had deceived ourselves into thinking we needed it.

We didn't need it. But I was blinded by my desire to have it.

"You know how you shoot a giant buck?" My good buddy Tom says. "By not shooting a big buck".

If you want something great, you gotta be willing to get rid of – or not take – something good. And just like passing on Tom's proverbial good buck, we sold our Hudson's Hope property and captured ourselves a giant buck – Freedom 35, which proved to be larger than life.

The Year Everything Happened

In June 2014, we sold our original off-grid homestead in Hudson's Hope and paid off all our outstanding debt. It felt amazing. I never looked back for a second. I was thirty-three and owned my own house and land debt-free. I had skills and opportunities to make money and a piece of land that cried out for all our creative efforts. There was so much we could still do and we had the financial freedom to prioritize however we wanted. At the same time, there was an even bigger business opportunity in the works.

No matter what, 2015 was gonna be crazy.

When we look back, Rose describes 2015 as the year that everything happened. The first year of our new life did not disappoint. The big project we took on was also a big improvement in some ways from Forest Kerr. This one was a large transmission system development, including multiple new and modified substations in the Peace Region of BC – the Dawson Creek Area Transmission (DCAT) Project. It was an easy day's drive from home and the whole family could come along. At first, we just moved into the Pomeroy Hotel in Chetwynd, where the girls quickly attracted attention and charmed the hotel manager. Developing friendships with the staff made the stay fun and paid dividends when we had to make a very unusual request later in our stay.

We eventually moved into a rented house, making Chetwynd our temporary new home away from home. We travelled back and forth, mostly on our own schedule, allowing us to maintain forward progress at work and on the homestead. The project required more personnel than we had and, with an estimated 15,000 hours in front of us, it was the perfect justification to hire our first full-time employees. Surprisingly, we aced our first hirings, welcoming a top notch P&C specialist and brilliant telecom trainee. With a few of my friends and sub-contractors, we were a formidable team of six.

The team was exceptionally competent and the project sailed along smoothly. I was having fun managing the work and keeping the guys happy while Rose wrestled with ramping up the logistics of a larger company. We were constantly learning about payroll, taxes, benefits and insurance, etc.

On the home front, we struggled with another winter of generous snowfall. A few years earlier, I purchased a 72-inch Berco Mac self-powered snowblower built for a UTV and modified it to fit on my Jeep Cherokee. It was pretty cool but was not suitable for the hills and curves of our long driveway. It blew shear pins like it held stock in mild steel. Next, we bought a great big hydraulic V-plow for our one-ton truck. Surely a giant plow on a giant truck ought to do it, we thought. It was excellent until the snow banks built up to six or seven feet, at which point the freshly plowed-up snow hit the banks and slid right back down onto the road. The result was an increasingly narrow trail resembling a luge track rather than a road.

The first year at Tipper Creek, a logging contractor kept the forest service road clear as they were working further down our road. All we had to do was clear our mile-long driveway. But for the next three years, there was no logging and no one else to keep the main road open.

It was time to up the ante. I started looking for a machine and found a super cool-looking thing at the Bobcat dealership: a Toolcat with a snowblower. It was used but still not cheap. At around fifty grand, it cost more than our entire property. We struck up a rent-to-own contract so we could try it out before fully committing. The Toolcat had a truck-style chassis, allowing it to drive and steer like a normal vehicle, but it also had rear-wheel steering, allowing it to turn on a dime. With a front loader attachment and high-flow hydraulics, it delivered forty-five horsepower to a six-foot-wide snowblower. If you've ever seen a three-year-old driving his toy truck around the living room, up walls, over chairs and smashing through his sister's doll house, he's actually driving a Toolcat.

The machine was every man's dream. It could just mosey along, blowing eight inches of snow off the road without breaking a sweat. With a steering wheel and gas pedal, it was encouragingly intuitive to operate. The older girls became competent operators almost immediately. The Toolcat was incredibly versatile on our property – moving dirt, logging, augering holes, ditching and brush mowing. It made itself indispensable. I've often told people that if you add a $50,000 Toolcat onto a $50,000 property, you get a one-million-dollar property.

By mid-March 2015, we knew we would be spending a good amount of time this year – and maybe next – in the Chetwynd area. So we bought a small four-acre property nearby. It was mostly sloped but faced south and had a beautiful view over the lake. At thirty thousand dollars, it seemed like a good investment.

More importantly, it would allow me to build a solar demonstration project. Since our first off-grid solar purchase in 2007, product availability had skyrocketed and prices had plummeted. This property was on the grid and I wanted to try a grid-connected system with net metering. Aaron and Elisabeth brought up the Toolcat and some culverts and helped us build an access road. Mostly timbered with mature poplar, it was a beautiful place to build a campfire and spend long spring evenings. We eventually built a pole barn and installed 10.5 kW of solar panels, each with its own microinverter. Just by looking at historical weather norms for the area, I estimated the system would generate about 12,000 kWh per year. Because there was no load, I accurately predicted it would pay me back about $1,200 a year.

For the remainder of 2015, we bounced between work on the DCAT project, the new solar property, and home. After years of just chipping away, we were finally crushing it on the homestead. Our shop had stood as a glorified carport for the last year and we finally sealed up all the walls and poured the concrete slabs with in-floor radiant-heat.

We experimented with two different types of structurally insulated panels (SIPs). On half the shop, we purchased very expensive, pre-manufactured panels with plywood and foam and, on the other half, we used straw bales plastered with mud. Offering the same structural profile, they couldn't be more different in their cost, installation, and final product.

The foam plywood panels cost about $8,000 and we used long screws to fasten them into the posts and beams. The installation took about a day. The final product was a flat, usable plywood surface inside and out.

The straw bales cost just a few hundred dollars and took a huge team of volunteers many days to install. The final product has an appealing earthy look, especially on the outside. But on the inside, the lumpy mud finish offered no support for shelving, etc., and shed a little bit every time it was touched. Both systems can be a win, depending on the application.

With the shop finally closed in, we were able to complete our final rainwater system, collecting off the shop roof into 900 gallons of storage.

Our little twelve-volt power system in the house was finally made obsolete by a 3,000 watt solar array on the shop roof and a 4,000 watt inverter/charger. First, I used a salvaged forty-eight volt industrial lead-acid battery bank but, within a couple of years, replaced it with lithium iron phosphate batteries. The new power system was a game changer and allowed us to have a regular fridge in the kitchen and a deep freezer. We also installed a regular AC water pump with a pressure tank on our rainwater system, allowing us to have running water in the shop. With pressurized water, we could force it through four stages of sediment filter down to 0.5 μm, which enabled us to eventually get a washing machine.

Even without any heat that first winter, the shop didn't drop lower than -12°C despite the coldest exterior temperatures hitting -40°C.

In the house, we finally built the remaining interior walls. For years there were curtains, and now, finally, there were actual walls. Except for the concrete floors, everything else in the house was various forms of wood. We had cedar wood posts topped with weathered spruce beams wrapped with fir plywood on the exterior walls. Exposed pine and spruce log rafters with more plywood made up the ceiling. In the kitchen, Rose's shelves, pantry and giant timber kitchen table were all made of various species of wood. The interior walls were much the same, so we added some homemade stucco and painted it white for contrast.

We purchased a few more LED lights – enough to walk around safely and look cozy but not enough to read a book with. Since the beginning of our off-grid life, we used our headlamps at night for task-specific lighting and still do even now.

Outside, we continued to clear land – cutting brush, felling trees, and letting the sunlight in. We cut in a few more small pens out of the thick conifers for our growing herd of Nubian milking goats. Rose built raised garden beds out of straw bales, planting lettuce, peas, cabbage, broccoli, and cauliflower directly into the bales with just a skiff of topsoil on top.

Without breaking the hard clay soil or pulling any roots, her straw bale garden produced fairly well. Once the garden was harvested, the bales had broken down into a thick mulch and were perfect for planting garlic in the fall. The garlic harvest that next year was outstanding.

Even in winter our cabin was warm and cozy

Perhaps the biggest development at our property that year was not even on our property. The portions of the river we had hunted the previous autumn were too good to not enjoy in the summer. In late winter, with a couple of crusty feet of snow on the ground, we snowshoed down and ribboned a pathway. It was over a mile with some very thick sections and some steep slopes but, together with Dave, Shoshanna, and the boys, we hacked it open. We were a wrecking crew of eleven, some cutting deadfall with the chainsaw, some dragging debris off to the side, one brush-sawing, a couple raking piles, while someone else followed up with a quad, proving the brand-new path. Walking and quadding it regularly quickly turned it into an idyllic park-like trail.

The gravel bar and beach at the far end were even better than we remembered. It was mostly cobble and gravel, but with pockets of soft sand to rival any holiday destination. A half-mile of river curled around the beach with a gradual shallow entrance on this side and a deeper channel on the far side. It was perfect for floating in a tube and some casual fishing. On the downstream end, it transitioned into a big slow pool. For the first time, we were able to hop in and try out the eddy current "washtub" nestled in the clay cliffs.

The water was perfectly warm – refreshing, but not the glacier-fed kind of refreshing. Fed from the large interior plateau of BC, it had the typical tannin-rich tea colour of the boreal forest but also an early freshet that peaked early in spring, allowing it to warm quickly in the summer. It became a staple summer activity and a favourite for friends and visitors.

After four years of slowly developing the homestead, the new private beach suddenly propelled it from "handyman special" into resort status.

The 2015 hunting and backcountry adventures also eclipsed all previous years. It started early in February when Abigail, just twelve, shot her first elk. She had a winter draw for cow elk and we went out once or twice a week when I could squeeze it into my work schedule. We had stomped way up a hill in deep snow where she killed it with a perfectly placed single shot with Rose's .243.

When we finally dragged the animal down to the road, we were met by a conservation officer (CO) and RCMP constable.

"Someone reported a poacher shooting on the road," stated the CO.

The RCMP officer immediately checked our guns and my firearm license and recognized we were not a risk.

"Well, there's no firearm offense here," he said, looking at the CO. "I'll leave you to it."

"We parked on the road a mile back, hiked up the hill and shot the elk on the ridge," I explained to the CO. "We were way off the road."

No one had seen us hunting or shooting the elk and there was obviously no evidence of poaching. It was a clear case of hunter harassment.

"Ok, let's see where you shot it," the CO insisted. We marched him back up the hill to prove that we didn't shoot it by the road. We wore him out before we even got back to the kill site and he tapped out. Back at the truck, he shook my hand and thanked me for raising my daughter right and teaching her how to hunt.

Legality and ethics aside, sometimes you have to shoot first and sort out the logistics later. In this case, we were still staying at the hotel in Chetwynd and, on our way there, wondered where in the world we would place an entire elk. Rose asked the hotel manager if she knew where we could find freezer space. As a hunter herself, she did not hesitate. "There's an empty deep freezer right in the breakfast room, you can use that," she offered.

"Are you sure?" Rose pressed.

"No problem," the manager insisted.

The four quarters, around three hundred pounds, fit perfectly and remained there undisturbed until we could cut it up. Each morning that week, I sat there looking around at all the unsuspecting strangers eating their scrambled eggs and hashbrowns, clueless to the entire body stuffed in the freezer just ten feet away.

It turned out to be Abigail's year, as she also shot a 48-inch bull moose in August.

Since we were in the area, we decided to go back to where our moose-hunting careers began and had Shawn join us in Hudson's Hope. Without Old Blue, we relied on three quads and two horses. We had purchased Mec, another Canadian, a few months after Nokee's running-away incident.

Mec had proven himself as a reliable mountain horse. A year earlier, Sarah and I packed Mec and rode him close to eighty kilometres up and down the rocky mountain foothills in the Muskwa-Kechika, chasing sheep and elk – bumping into a few grizzlies and bison as well. Jay and his oldest daughter, Sierra, accompanied us with their Arabian. It was a fantastic adventure and proved the capacity of the horse. At one point, Sarah had ridden him into a muskeg bog and he quickly sank to his chest. She didn't even have to hop off – she just stepped off, holding the reins. Thankfully, Mec remained calm and did not flail. A more flighty horse would've thrashed around and sunk past his ears within a matter of minutes. We swung his neck around and pointed him back to hard ground. With Sarah pulling on the reins and me pulling on his saddle with the lead rope, he plowed his way back to the surface.

This year, it was the whole family, and we were looking for any bull moose. Opening morning, Abby and I sneaked up a little creek bottom into Brutus Valley, where we'd seen many monster moose in the past. Sure enough, on the steep far hillside, we spotted a great big bull about two hundred and eighty yards away. Abby shot first with the .243 and I followed up with the 7mm. He fell down and never got up.

We ran back down to camp, gathered the whole crew, took quads and horses the long way around that led up an old road on the far side of the valley. Leaving the quad trail, we led the horses over the lip of the ridge and straight down about four hundred yards to the moose. The knee-high buckbrush gave way to twenty-foot alders. With the annual crush of deep snow, each alder stem – maybe three to four inches across at the base – was laid flat to the ground, aiming downhill. Travelling five or ten feet within just inches of the ground, they would make a quick sweep up toward the sky where the stems were one to two inches across. Strong and flexible, we could easily walk on the stems and push them down in front of us going downhill. But trying to go back up against them was like fighting a Chinese finger trap. The harder we pushed, the taller they stood up. Each one, like the limb of a bow, resisted our movement and by the time we made it a few feet, we were fighting a dozen of them.

There were only two options. One was to angle downhill and find a seam of conifer in the alder, then scrape our way up through the spruce. The second was to literally crawl under the creeping alders, pushing them over the top of our bodies.

I had no idea how we were going to get back up with a moose.

Our two bullet holes were both side-by-side in the kill zone. We struggled to carve up the giant beast on such a steep slope and had to get very technical to load the quarters on the horse. It seemed like we had bitten off more than we could chew with our first attempt at packing meat with our horses. We could not even stand without holding onto something, and to make our way back up seemed hopeless.

Uncle Dave wasn't there, but I could hear him say, "We're not gonna pack it out just standing here."

I felt like Peter in the boat when Jesus asked him to step out; it was going to take a miracle to go back up the hill. But, like Peter, I took a step in that direction and what happened next blew my mind. At first, I aimed toward the nearest spruce tree, thinking I would crawl and force my way through the branches. I did that and made a whopping five feet of progress – not even far enough to stretch the lead rope. Nokee just stood there confused while I attempted to move on, wriggling and writhing under a mat of alder. When I finally got to the end of the lead rope, I turned to Nokee, gave it a tug and made a "Clk, Clk."

I didn't know what I expected him to do. I decided to let him figure it out. And that's when I learned what 1,600 pounds of muscular beast can do. He put his head down and charged straight up with total disregard. Like a war horse, he lunged and crushed everything in his way. His feet dug deep and threw clumps of soil in the air.

I had unlocked monster mode and was frozen in disbelief. Our mild-mannered horse had just erupted into a raging machine and, with the lead rope clutched instinctively in my fist, I was sucked into the action.

"Whooooaaa, whoooaaa, whooaaa!" I yelled.

Rose and the girls yelled after me but it was indistinguishable. There was no possible path straight up the hill, and yet that's exactly where we went. Without any strategy or deviation, Nokee pounded through whatever got in his way. Like a fallen water skier clutching foolishly to the tow rope, I plowed through the rhubarb head-first behind him. My survival instincts were focused almost entirely on avoiding the hooves, and I was only distantly aware of the abuse I was receiving from the ground and the alder boughs.

I tried but failed to get my feet under me. Mercifully, Nokee came to an abrupt stop after fifty yards. He wanted confirmation that we were still going the right way, but I did not give it.

"Are you OK?" I could hear Rose yelling but, in trepidation, I remained quiet.

Nokee's chest was heaving so hard I thought it might tear open and I was still checking myself for injuries. I distinctly remember brushing debris off my face and feeling my eye sockets to make sure I still had both eyeballs. Nokee had solved the problem of how we'd get out but had created a new problem for me: how do I not get killed? The only possibility was if I also took my game to one hundred.

Before the adrenaline could dissipate, I prepared for battle and gave Nokee the word.

"Clk, Clk. Let's go!"

He burst into four-low and it was immediate hand-to-hand combat. I defended myself from the pummeling alders by attacking the weakest point, splitting their defenses and throwing myself against the wall. The best defense is a strong offence, or something like that. We emerged on the crest of the hill, beaten up, but victorious. Rose had started out behind me, leading Mec. We had yelled frantically back and forth a few times, but now Mec, bearing his load of meat, came bursting out of the brush by himself. Following our lead, he had lost Rose at some point and just kept going. He came and stood patiently with me and Nokee as Shawn and the girls slowly trickled out of the bushes.

"Rose..... Rose?" I called, wondering if she was gone forever.

Eventually, she stumbled into the open and answered, "Yes, I'm here."

If I hadn't experienced it, I would not have believed the explosive power these gentle animals possessed. I was grateful for the strength, impressed that even when we had clearly lost control, they never got flighty or panicked. A lesser horse would have run off like a three-year-old in an airport and we'd still be looking for them.

The whole family dwarfed by Abigail's moose

Wandering in the Wilderness

The following month, we would push the limits of our family and the horses even more. We found a small guide outfitter territory for sale in the Big Creek area of the Chilcotin. The under-utilized, five-hundred square mile area lay on the northeastern slope of the Coastal Mountains with no motorized vehicle access. We "planned" a weeklong trip along with Bev, Jay, their girls, Sierra, Cora, and River, and their horses, Lady and Tank.

The plan consisted of where we would park the truck and roughly how long we'd be gone. It did not include exactly where we would go or how we would get there. We had looked at some old maps that showed the location of the main camp, which became our first target. From there, we hoped to make our way to the high-elevation southern parts by the Dil-Dil Plateau and Nadila Lake. I had studied the area on Google Earth and, fortunately, the images cached on my phone. That satellite imagery and the little blue dot showing our location would prove invaluable.

We set out, leaving our trucks and horse trailers in a clearcut, and headed west. I grew up – whether it was with Uncle Dave or Auntie Joy –

with pre-planned routes that always included documented trails. I was used to going off-trail when hunting and I loved it, but I'd never headed out for a week without a real plan. Jay, however, was born for this. Jay comes fully alive when he has no idea what's coming next.

The first full day we followed some cattle trails through range land down to Big Creek and found a crossing. On the other side, we made camp by a little moosey-looking lake and started dinner. Rose had, of course, planned and packed all the food and so I was shocked to see a big heavy bundle of carrots come out of the saddlebag for dinner. I thought it was some sort of joke.

"Who brings five pounds of carrots on a backcountry expedition? Rose, did you pack five pounds of carrots?"

I was being logical, which is the same as being a jerk, and she got embarrassed. With two horses carrying food and gear for seven people for seven days, we did not have capacity for five pounds of carrots. Each one of us – even all the girls – were carrying backpacks with our own sleeping bags, clothing, and personal items in them. We had no extra room and yet we somehow fit fresh carrots, potatoes, onions and meat as well.

While secretly looking forward to the hearty dinner, I loudly admonished Rose, "Are you trying to kill the poor horses? Is there any food in there for the next seven days?"

Rose defended herself in dialogue but the girls came physically to her defense, immediately and aggressively. Like a flock of geese they came at me from all sides, hitting and gouging and perfectly exacting justice.

Jay came over to see what the fuss was about and, after hearing the explanation, he laughed and named our little body of water "Carrot Lake."

The next morning, we were jolted awake by a shaking of the tent and a loud whisper.

"Hey Jeff, get up!"

"What. What?" I responded in confusion.

"There's a huge moose right here," Jay replied.

We all squirted out of the tent one by one, like the birth of a litter of puppies, wriggling into our clothes and our boots. Jay was hurrying us along, waving us to the edge of the clearing.

"Over there," he motioned.

We quietly inched our way out of the timber and into the tall swamp grass. There, just one hundred yards out, was one of the biggest, most magnificent bull moose I've ever seen.

The sky was turning robin-egg blue as the sun hit the tips of the spruce trees. It was only our second day and we were already mesmerized by our first magical moment in the wilderness. The bull was in no hurry and stood for a long time in the clearing as a monument to creation. You can spend ages in the woods without having such a beautiful, peaceful interaction with a wild, awe-inspiring creature. I was thrilled for all the girls to be a part of it.

Feeling like champions, we packed camp, not realizing the winning had just started. We marched onward, blue sky overhead, sun blazing in splendour and the forest emitting the sweet, pine-resin smell of summer vacation. The girls were all finding their grooves, bouncing around under their packs and taking turns leading the horses.

The timber grew thicker and thicker and soon we found ourselves stuck in a whole forest of pecker poles. The horses were nearly six feet wide with their panniers and it became a constant challenge to find a route.

Mec was trained to pull a cart or plow, which is great if he's pulling a cart or a plow. A beast of burden, when feeling resistance, instinctively digs deeper and pushes harder. If we led Mec into a tight spot and wedged him in with the saddlebags, he would just drop a gear and push through, saddlebags be damned.

It was tedious, but we had scouts go ahead and then specify the exact path for the horses to follow. After ten miles in the timber, we emerged into a clearing and followed a web of natural meadows to the main camp.

The old outfitter camp was a picture of rustic beauty tinged with sad neglect. It was made up of three small cabins, corrals, a little barn, and horseshoeing shed all tucked into the trees with a view across the meadow and up to the snowy peaks. Apparently abandoned for decades, the cabins reeked of urine and were boldly guarded by pack rats. Mattresses were shredded into giant fluffy heaps, weighed down with a mattress topper of rodent poo. The kitchen in the main camp still had dishes on the table and a fry pan on the stove. There were shelves with canned food and even a glass jar with sour keys in it. How did they know I loved sour keys? They were rock-hard, but I ate one anyway.

We flipped through a photo album of all the successful moose hunts from the distant past and read through the names in the visitor book. The camp was probably beyond salvageable but was, nonetheless, inspiring.

From outside came the beating of grouse wings and we went to investigate. Sierra was lining up the .22 with Jay's help. She popped it. She ran up, claiming her plump chicken prize and rustling a few more birds from their hiding spot in the process. Jay invited Sarah over, and she bagged her own dinner.

With every shot, three more grouse came out and waddled around, advertising themselves. Jay patiently helped each kid – all seven – harvest a bird. Sometimes bravely staring down five or six bullets, each fool hen sacrificed itself for the cause. The kids were pumped, like a team that had just won the Stanley Cup. They held their trophies high.

Our oldest, Sarah, was fourteen and Keziah, our youngest, was just about six. At that age, I could've only dreamed about such an adventure and it warmed my heart to see them soaking it up. We pitched our camp, ate a huge meal of fresh meat, and then sat around a campfire talking about our next goal: the mountains.

The following day was difficult. We gained a lot of elevation over rough terrain and struggled with heat and exhaustion. We didn't know where our next water source would be, and we were conserving what we had. The girls had a hard time leading the horses up the steep, rocky trail and would occasionally get knocked over by the knee or nose of a horse. There were tears and periods of silence but, eventually, the whole crew of us emerged at Twin Lakes.

A huge grassland surrounded two lakes, broad and shallow. They were more like fifty-acre puddles. There was fresh grizzly scat along the trail, so we peeled off and tucked in along the timber 400 yards away.

That evening, Rose and I left the kids at camp with Bev and Jay and sneaked back out to a vantage point overlooking the lakes. We watched three bull moose make their way to the edge of the water and push each other around. The lakes were glass-flat and perfectly reflected the snowy Coastal Mountains to the southwest. The buckbrush and willows were wearing their vibrant fall reds and yellows, and the cloudless sky was vibrant blue.

"Rose," I whispered, "Let's move out here."

She squeezed my hand in silent reply and we sat still, soaking it in. This whole trip had been a tour through a living, breathing calendar of stunning scenery and it wasn't over yet.

The next day we pushed on, the landscape transforming from thick timber to rocky alpine. We traversed the vast, rolling and folded terrain and were offered a picture-worthy mountain view every time we lifted our eyes. We were approaching Ram Mountain, a lonely bald outcrop that seemingly held back the glacier-fed waters of Nadila Lake. Still miles away, we were thrilled to find small seeping springs along the slope. We were out of water and the miles of climbing had us dripping with sweat. Sometimes just a little puddle the size of a salad bowl but filled with icy-cold water bubbling from underneath, these were well-known by the local wildlife and were marked by game trails.

Rested and with water bottles full, we pushed up and over the saddle separating us from the lake. Clutched in the jagged grip of the mountains and shaded by the tallest peaks, it seemed like the perfect place to build a little cabin and live the rest of my life. Reminiscent of Dick Proenneke's famous homestead, it was a living postcard.

As we all got closer, our excitement and imagination grew. All composure was lost. We abandoned our rank and file and scattered as each adult and child hurried to discover their own personal paradise along the lake.

Jay and I instinctively ransacked our packs for the little collapsible fishing rods we each brought. While some girls scrambled to explore the shoreline, others were casting and catching explosive little mountain rainbow trout.

It was about then we noticed the shift in weather. Clouds rolled in, the temperature dropped, and the reality of our high-elevation mountain position took hold. We were all still drenched with sweat and, as the breeze picked up, were forced to take shelter. With nothing but rocks and brush, the girls found a little low spot and managed to shove the tent in between some mountain birch. They donned all their warm gear, lit a fire and cooked up a couple of little trout. The rainbows-and-unicorns part of the trip was over. It was time for the dragons.

On the Dil-Dil plateau with Nokee and Mec

Little kids will amaze you if you let them. If you can do it, they can do it. Obviously, given their size and strength difference, there are limitations, but that's not the point. If you have a positive attitude, they'll generally have a positive attitude. If you think you can do it, they'll think they can do it. If you're terrified, they will be terrified.

The next day would test us all and ultimately prove my point. It was cold and we were exposed, with minimal shelter. To compound matters, clouds were building and it looked like snow was coming. We were navigating mostly by sight and, if it socked in, navigation could prove nightmarish. In this steep rocky terrain, any real depth of snow would obscure our footing and reduce our traction. We were in imminent danger. We were at 2,800 metres elevation and the fastest way to lose elevation was to cross the Dil-Dil Plateau and drop all the way down to Big Creek.

With no trail to follow and no time to spare, we broke camp and headed east. We climbed a couple of hundred meters out of the Nadila Lake basin and, for the first time, saw the length and breadth of the Dil-Dil. From this distance, we could see no defining features and just aimed for the middle. It began to snow as we crossed the great expanse of boulders and low brush, spotted with oversized puddles and small marshes. The snowflakes were giant and wet and they soaked us through. Thankfully the temperature was right around freezing, so the snow struggled to accumulate.

We were picking our way through with our heads down when a covey of ptarmigan burst from their hiding place and sailed away. Mec was carrying Grandad's .410 in a scabbard this whole trip for exactly this purpose. Already turning their winter white, they blended in perfectly with the brush and patchy snow. But we had a lot of eyes on our team and I tracked down and shot two of them. I was thrilled to shoot my first ptarmigan, but we had no time to celebrate or prep them. I just threw them in the saddlebag and we kept moving.

Sodden and weary, the ten miles of zigzagging and weaving obstacles was pushing our crew to the limit. At a lower altitude, we would've just pitched camp, but we couldn't risk staying at 2,000 metres.

It was late afternoon as we neared the edge of the plateau. It did not gradually slope to the river valley below as we had hoped. We edged our way to the top of a crumbly rock cliff, scanning the expanse ninety feet below. From what we could see, this cliff continued for miles in either direction. Without knowing the lay of the land, we had navigated straight into the middle of a topographical conundrum. We didn't have time to explore for miles in each direction and we knew the persistent snowfall would pile up overnight.

We had to go down.

It was Jay's time to shine. He's half mountain goat, which is exactly what we needed. He bounded along the rim, looking for any possible path. To me, it seemed totally impossible.

"Check this out, man!" he called.

We examined the cliff together. There was a cleft in the basalt, and he jumped down over the edge and scrambled down the ten-inch-wide rock ledge at a forty-five degree angle. He was using his hands most of the way. The cleft made a sharp switchback onto loose rock, giving way below his feet. Moving quickly, he scampered down the loose rock – sometimes on his feet, sometimes on his butt.

"Whoa, you think we can do that with the horses?" I asked.

I knew we could get down but, for the horses, it just seemed crazy.

"I don't know," Jay admitted, climbing back up. "What other option do we have?"

With no competing proposal, Jay and Lady took the lead to find out. Jay jumped down again, yarding on Lady's lead rope. She stood stubbornly at the top, tentatively pawing over the edge as if to prove that there was nothing for her to walk on. Jay doubled down, tugging her nose down to the edge.

"C'mon Lady! Clk, clk," he coaxed.

She hesitated, offering Jay a chance to change his mind. When he didn't back down, she faithfully made the jump. Her hooves landed on the sheer rock but didn't have the traction to hold her on such a steep grade. She slid, scrambling frantically, and fell, knocking Jay off the ledge. Rose gasped and covered her eyes.

I could hear one of the girls utter a tortured, "Oh, no!"

Lady's rump and shoulder scraped along the rock face and she ground to a halt in a twisted heap. Jay righted himself in the loose rock at the bottom of the face and ran back up to grab the rope just as Lady clawed to her feet. She was quivering and had a cut on her hip. But with Jay back in charge, she slid and scrambled down to safety.

The Canadians, bred for rough terrain, had the advantage of following Lady's lead. Their whole bodies shook as muscles strained to hold their weight. They hit the switchback hard, plowing into the loose rock. After pulling them over the edge, we just got out of the way and tried to stay right-side up.

With the horses down, I immediately turned my focus to the miles we still had ahead of us.

Rose scrambled down with the little girls in tow and stood beside me. The adrenaline subverted the girls' exhaustion and, without celebrating, we dragged ourselves toward the timber.

"We've got to get to the creek bottom," I encouraged. "Let's do this."

It was already getting dark in the forested slopes as we pushed through the deadfall to Big Creek. We blitzed the crossing, barely even slowing down. On the far side, we threw our packs down and climbed into a patch of thick spruce to stay dry. We tied up the horses, pitched the tent, and piled in.

We woke to an uncommon darkness and an eerie silence. A blanket of snow on the tent emulated our own mass grave. After our harrowing march the day before, it almost seemed appropriate. We shook the snow off and popped outside to take in our new winter surroundings.

It was beautiful in the valley next to our bubbling mountain creek. But back up on the mountain, it would've been deadly.

The girls were sick of instant porridge packages and threatened to mutiny. It was Rose who sunk a dagger into my authority, stating

"No. I am not eating another instant porridge!"

The girls piled on like a pandemonium of parrots.

"We're not eating porridge. No way!"

"Fine, we can eat the ptarmigan," I taunted, pointing to the still feathery lumps of meat.

In an act of defiance and without hesitation, Rose agreed.

"Ok, and we'll have it with Mr. Noodle soup."

For some reason, that "brown soup," as the girls called it, and barely cooked wild chicken for breakfast has remained a backcountry benchmark.

A relatively easy two days' march along the river put us back at the truck. I had finally taken my whole family on an epic adventure in the mountains. It felt like now they knew what the inside of my heart looked like. The girls had seen so much and done so many new things. They had worked harder and been stronger than they'd ever been. Once you've done really hard things, nobody can tell you that you can't.

And Rose had survived. The trip had certainly been most difficult for her. I had experience and confidence (arrogance?) and the kids were fueled by blissful ignorance. Her protective mother's heart was tested and, yet, for the sake of her family, she overcame and even enjoyed it.

In 2015, I was an all-star in my rookie season. It was our first year debt-free and we scored a lot of goals. We managed to pack in years' worth of business, homestead building, and outdoor adventures. While I was away at work, Rose had overcome the death of her trusty bear dog, Hunter, and an injury to Nokee that left him blind in one eye. Sarah shot a nice 4x4 mule deer buck in November and, in December, we travelled around Guatemala and spearfished Tobacco Caye in Belize.

The choice to sell our perfectly wonderful marshmallow house and dream property for a debt-free life in our modest little off-grid homestead had brought us to this point. It seemed like we had it all, but life is not a zero-sum game and, somehow, it was about to get even better.

I wanted to share homestead and hunting stories with friends and family and, in the winter months of early 2016, I started writing a blog and uploading short videos. It was fun and I did it for a few months, but I was clearly most passionate about making the videos. I sought advice from a friend about uploading the videos to my blog page and he suggested I start a YouTube channel instead. I had never seen YouTube before.

Relieved to not have to write anymore, I happily filmed and edited and uploaded cheesy little family videos. They were fun to make and the kids loved to watch them. I had no idea then that I would slowly and painfully improve our videos and still be making them nine years later. At the time, it was just a little creative outlet in between managing projects at work and development at home.

The DCAT project slowly drew to a close in 2016 and left me looking for work for the team. In June, I reached out to an engineering firm that I knew did work with BC Hydro. I called to offer the support of our field services team but received a much bigger proposition in return.

Freedom 35 and My Double Life

I was thirty-five years old, and it seemed we had successfully executed our Freedom 35 plan. But had we? And if so, what comes after Freedom 35? The plan had been to own a productive homestead debt-free. In the spring of that year, Rose and I evaluated that simple objective versus our actual state of affairs. We certainly owned a homestead debt-free. But was it productive?

By productive, did we mean self-sufficient? Self-sufficiency is the mantra of many off-gridders, homesteaders, and preppers these days, but it was just the natural condition for many people one hundred years ago. I think the allure of self-sufficiency goes hand-in-hand with our common desire to get out of the rat race - the idea that, if we can take care of ourselves, we won't be a slave to the bank, our jobs and all the expectations of a modern life in the city.

But what is self-sufficiency? Is it the ultimate goal, a means to an end, or some combination of both? Self-sufficiency means you will need to perform every kind of work. It will certainly involve home construction or repair, growing, gathering, preserving and preparing food. It will include all sorts of conditions and will often be dirty, messy and physically laborious, guaranteeing a lot of time spent outdoors. If that's part of your goal – if that's the work you want to do – then self-sufficiency is at least part of your end goal.

Our family has found this work is often rewarding, peaceful, healthy and outweighs a lot of the negative aspects. But how self-sufficient do you need to be? Complete self-sufficiency means you wouldn't buy or sell anything. You'd have to smelt your own iron, harvest your own sea salt, stitch your own clothes using a bone shard for a needle and sinew for thread from the hide of a deer you killed with a stick.

You would have to go one hundred percent caveman.

When pressed, most people aspiring for self-sufficiency will admit that they are actually aspiring to increase their self-sufficiency. If that's your goal then, at some point, you need to ask yourself, how far do I want to go – full caveman or just half caveman?

Having made so much progress on this ourselves, we got down into the weeds and tried to sort these details out. On one hand, I loved learning new skills and if I could actually smelt my own iron effectively, I would do it. Lord knows I would certainly be happy hunting bison with a spear while wearing a deerskin loincloth. But when you consider the people who actually did that – the Canadian Métis in the seventeenth and eighteenth centuries, for example – they were actually very reliant on their community. You can't hunt bison on the plains with primitive tools by yourself.

Similarly, do we truly want to exclude other people from our lives or would we prefer a strong community to be part of? We came to realize that we wanted to be more self-sufficient, more reliant on friends, family and neighbours and less reliant on governments and giant corporations.

Another argument for self-sufficiency is the improved chances of survival in a post-apocalyptic scenario. Well, I've never considered myself a prepper, but I do kind of like the idea of surviving. This might make your goals tricky because now you have to consider how bad the singularity is and how long you want to survive for. You could obviously write a whole book on this topic, and I'm not that author, but suffice it to say the requirements to sustain an off-grid homestead naturally tend to protect against disruptions in the outside world.

So, do you want to survive? Do you want to be a caveman? Do you want to quit your 9-to-5? Do you want maximum personal liberty? And how much comfort, convenience and community are you willing to sacrifice? While it's probably impossible to quantify the exact proportions of your perfect Venn diagram, I think we can rely on the simple urge for more or less of any of those things and move in that direction.

In 2016, we still wanted a bit more self-sufficiency. We wanted bigger gardens. We wanted to make our own milk, butter and cheese. And I wanted to reduce the stress and amount of time Rose and I were investing in our business. We were considering our options when that simple business call took an unexpected turn.

BBA, a national engineering firm, offered to bulk-hire my whole team, promote me to partner and give me the role of Director of Field Services. It would keep me engaged and give me a technical and managerial challenge while removing the stress of business ownership. It would open up a lot of new opportunities for our team of guys and keep them busy. My compensation would be great, but would lack the perks and profit I was currently generating. Rose, who had been working nearly full-time administrating, would retire and fully focus on the homestead and homeschool.

It would be the end of our traveling family circus, always on the move from homestead to our next project and back again. I remember camping at Okanagan Lake Provincial Campsite in July with my extended family while negotiating the deal. Sitting by the lake in our lawn chairs, taking in the view and the sunshine, we made our decision. It made sense on paper and, more than that, it was a shiny new thing that captured my imagination. Imagine me, being the Director of..... ANYTHING!?

In September, for the first time in five years, I became a full-time employee again. My first job was to integrate my field services team into the Vancouver office and then establish a nationwide field services department within BBA. As a privately owned engineering consultancy with eight hundred employees, it was a large but not massive company. It had credibility, with thirty-eight years of history, and had grown steadily from an office in Mont-Saint-Hilaire, Quebec, to six regional offices across Canada.

The ownership style was unique and I think it was structured very well. They maintained a ratio of partners to employees of about one to seven. On average, any project that required seven or more people would include a partner, ensuring the highest level of commitment and vested interest in project success. They also had a no-fat-cat policy – every partner had to be a full-time employee and no single partner could own more than five percent. With this model, no small group of partners could monopolize ownership decisions. The resulting partnership felt and operated like a team of peers instead of lords and serfs. Each partner was subject to annual performance reviews and, based purely on merit, either increased or decreased their number of shares.

BBA was then, and remains today, a successful example of pure meritocracy. BBA historically provided a lot of services in the field, testing, commissioning, and troubleshooting, but never had a devoted field services team. Their goal in assimilating our team was to offer technical field services to their mining, utility, and industrial clients, complementary to their engineering services.

It shaped up to be a dream job. I had almost full autonomy to pursue our business goals the way I saw fit. I defined my own work hours, workdays and work location. I traveled to the Calgary, Toronto, and Montreal offices occasionally and the Vancouver office every couple of weeks. The rest of the time I worked from home.

Of course, we had no office in our tiny house, but I didn't need much of an office. Each morning, I set up in the brightly-coloured Guatemalan hammock in our living room with my laptop and cell phone. Rose would bring me tea and then breakfast and I would write proposals, send emails, phone clients, and, of course, sit on Zoom.

I bought a few nice, collared shirts, a wool sportcoat and a couple of pairs of brand-new jeans. When I arrived at a downtown office, I fit right in with the engineering crowd but it often felt like I was wearing a disguise. We would grab a meeting room and discuss technical project specifics and business strategy. We would take clients out to fancy restaurants and pitch them our proposals. I would stay in expensive downtown hotels and stock them with snacks from super-expensive, trendy little organic grocery stores. A few of the partners owned sailboats and one even had a full-length fur coat and drove a Jaguar.

I felt like Donnie Brasco infiltrating the mafia, except the gangsters were hardworking, honest men. I was having a ton of fun, but I was certainly out of place. I wasn't against fur coats – it's just that I would rather hunt the jaguar and make the fur coat myself. I would sometimes sit smugly in a meeting, scanning all the suits around me, knowing none of them lived in a cabin in the woods.

It wasn't until about six months in, sitting for breakfast at the Château with a senior partner, that my lifestyle came up for discussion. He didn't quite comprehend the whole off-grid homestead idea and peppered me with questions to confirm exactly what I meant.

"What do you have for electricity?" he asked.

"Solar power," I replied.

"What about water?"

"Rainwater."

"But you have indoor plumbing and running hot water and all that?" he asked presumptuously.

"No, we just carry water to the house in jugs and heat it up on the stove."

He looked at me as if I was trying to sell him a used car, with an ever-deepening furrowed brow.

"Tell me more about that," he challenged, suddenly a little businesslike.

And so I told him about my Freedom 35 plan and how we live simply and debt-free. Like seeing the sunrise for the first time, his expression slowly transitioned to one of dismay, but also reflecting great concern.

"You have no debt," he said, matter-of-fact.

"No."

"No mortgage?" he pressed.

"No," I answered, apologetically.

"No car payment? Nothing? No debt at all?"

He was almost begging me. I shook my head.

He cocked his head and leaned in just a little. He squinted at me in an attempt to detect even a microscopic level of dishonesty. With more than a hint of annoyance, he stared right at me and managed only a single profanity. I snuck a quick glance at the other patrons in the rather quiet restaurant and then returned his stare with a poker face.

"We're gonna lose you!" he continued.

"What do you mean? Why would you lose me?" I asked honestly.

"Well, why would you stay? If you've got no debt, you can do whatever you want," he offered, looking visibly defeated.

I couldn't help but slowly break into a smile and, without confirming or denying, I replied, "I'm here for fun."

I found it very telling that, as an executive, he was keenly aware of the debt trap and how it benefited the company. As a blue-collar worker, I understood personal debt to be our own fault, but it is also easy to imagine it as a conspiracy against the middle class. Now, like a foster kid in the managerial class, I discovered they knew all about it. Using that knowledge hardly implicates them, but it does beg the question: What about the next level up – the elite class? Is it possible they are committed to keeping the rats in the race?

Regardless, we had found our way out and were happy to show others the way. In the meantime, I still had a double life to lead.

I was fairly diligent at keeping my headphones muted while on conference calls. Rose was often making a racket with cast-iron on the cookstove, so I was extra diligent. For quite a few months, I held hundreds of conversations without a hitch. It finally occurred to me that, with so much time spent on the phone and in meetings, I should probably get something better than the cheap earbuds I was using. The new Bose headphones I purchased were so comfortable and had great audio. The microphone, it turns out, was also very high-quality and picked up all the sounds.

I'd been on conference calls for hundreds of hours without any problems, but suddenly people were mentioning weird background noises. I was aware of how sensitive it was but did not yet understand the extent of the problem. Within just a few days of getting the new headphones, we had our monthly executive roundtable. As a Director of a business line, I reported challenges, opportunities, and significant actions taken in the previous month. I only had to talk for two minutes and then listen for another twenty-eight, so I only needed it really quiet for two minutes.

The girls were outside milking and, as it came to my turn, I motioned for Rose to keep it down. I unmuted, began speaking, and immediately knew I was in trouble.

"Hi, this month we won two contracts with a new client...uh..."

I could hear kids yelling outside and they were coming closer. Even as I continued talking, it became difficult to focus as my subconscious attempted to decipher the reason for the approaching melee.

"Uh... the new client is...."

My words became disjointed, the pauses between words grew, and then, as the kids ran up to the front door, I just stopped. Expecting them to burst through the door, I was surprised when they ran right by our large, south-facing windows, chasing two very excited goats. Rose had run to the door to intercept them and motioned for the girls to be quiet, but there was no stopping the goats –

"Maaaaaa, Maaaaa!" they bleated, followed by more goat bleats fading into the distance "Maaaaaaaa, Maaaaaaaaa!"

Then silence.

I was silent. Rose was silent. The kids were silent. And the entire executive team was silent. It was a moment of silence.

Rose was staring at me with wide eyes, clenched teeth and eyebrows raised in a hopeful, maybe-no-one-heard-that kind of way.

"Ahhheemm," came the clearing of a director's throat, followed by more awkward silence.

"Uhm, I think I heard... goats," someone suggested. "Whoever has the goats, maybe you could mute your microphone."

It was the first time my secret lives intersected and raised some eyebrows, but it would not be the last.

It was spring, and we had been invited to hunt feral sheep on one of BC's many coastal islands. I was working in Vancouver for the week, so we drove down with the whole family and moved into the company apartment downtown. It was a nice, modern two-bedroom suite on the seventeenth floor of the Pointe Claire Building in Coal Harbour.

Just like old times, Rose and the girls did homeschool during the day while I was at work and, in the evenings, we explored downtown and the waterfront. A family of seven sticks out anywhere these days, especially downtown, and especially with five beautiful girls. Even more so when they're wearing camo and gumboots. We normally didn't mind extra attention, but we had rifles along and had to smuggle them in and out of the apartment. We used our biggest backpacks, stuffed the rifles in and pulled jackets and blankets over the protruding barrels. It looked ridiculous, but clearly no one down here would suspect a couple of preteen girls of sneaking firearms. As it turned out, the guns were the easy part.

Wrapping up the work week, we headed for the ferries and made our way to the island. For decades, it had been overrun by feral sheep that were destroying native vegetation and out-competing the wild blacktail deer. The sheep were a contentious issue amongst locals. Some wanted them eradicated and others threatened violence to anyone who would harm them. What the locals almost unanimously agreed on is that mainlanders are not welcome, no matter what. From the moment we climbed on the small twenty-passenger ferry that serviced the island, we were met with suspicious stares from the locals.

The sheep look like regular woolly domestic sheep and are readily visible anywhere with low vegetation. We shot four throughout the weekend, hung the meat outside in the cool coastal air, and salted and packed the hides. Our covert operation was now in the hide-the-body stage. Our plan? Put the bodies in a suitcase. No one would ever suspect Sarah of walking around with a suitcase full of bodies. As long as they didn't try to lift it or smell it, we'd be fine.

With their backpacks full of camo, rifles, and ammo, and dragging a sixty-pound suitcase of meat, I followed the girls along the dock. I helped Sarah heave the suitcase over the ramp and into the forty-foot aluminum ferry. It's hard to conceal effort, and so, just like a guilty person would, I quickly glanced around to see if any of the other passengers had noticed. The locals all looked at us judgmentally anyway, so I couldn't tell the difference.

Crossing the strait, we pounded through heavy seas; as the suitcase jostled about, I eyed it continuously in fear blood would start seeping through. I had a feeling the islanders needed very little justification to throw us overboard.

We made it off the boat, but we still had to make it to our apartment. Steering the suitcase along the road, I could feel the weight of it every time it jarred at a joint in the sidewalk. I'm surprised the concierge didn't phone us in. We looked dishevelled, smelled homeless, were carrying suspicious-shaped backpacks and dragging a suitcase that clearly contained bodies.

And why were there so many little girls? Had we kidnapped them? He just looked up, gave us a robotic, "Good evening," and went back to his crossword. We beelined for the elevators.

"Just about there," I whispered to Rose.

It's an interesting thing in cities. Most people seem to actively ignore their surroundings. They don't make eye contact. They don't tip their hat or nod. They just pretend to be alone.

If it were me and I saw us all climbing in an elevator, I would wait for the next one. But some lady, in the city way, didn't look at us. She just crammed in – her and her dog. Rose saw the dog, shot me a worried glance, and quietly said, "Oh, dear."

The dog sniffed us and all our gear and was clearly overwhelmed. Rose was holding my hand and crushed it as the dog came to the suitcase, sniffed it and cocked his head. Any self-respecting farm dog would have immediately discerned the meat and sniffed and pawed or whined and barked, drawing attention to our bag of bodies. But this dog was a city dog so distantly removed from its wild ancestors that it could not make sense of the hunting potpourri we provided.

After a long nervous ride in the hotbox to the seventeenth floor, we rushed to our room. I pushed all the troops through the door, closed it behind me and laughed.

"Ha! We did it. We did it, girls!"

Pouring into the living room, we discovered it was essentially a fishbowl. Dark outside, but with our lights on inside, we were on display for all the other downtown towers. Rose looked out with renewed concern.

"There is no privacy in here," she exclaimed.

But I was feeling confident now.

"It's like pay-per-view," I said. "What are they gonna do?"

We opened up the suitcase on the living room floor and, one by one, pulled out pieces of meat and laid them across the dining room table. Within a few hours, it was all cut and wrapped and packed in the freezer. Twelve hours after that I was all dolled up at a business luncheon. Part bushman, part mountain man and part businessman.

I was leading a secret life and loving all of it.

The Open Ocean

For the first time since we first moved to the Tipper Creek property, I was home most of the year. Our gardens grew and we fought back the brush and dug out the old stumps in front of our house. The cedar siding on the house had turned to a rich reddish-brown with bleached streaks at the bottom wherever water regularly splashed. We improved the soil on the roof and, each spring, the grass grew thick and rich. When it faded in the summer, a sprawling mat of succulents kept it green.

Sarah bought a forest-green '94 Jeep Cherokee from my buddy, Tom, in anticipation of her sixteenth birthday. I had told the girls I would help them fix up any old vehicle they bought, as long as it was an old Jeep Cherokee.

We cleared a patch of poplars up on the hill, borrowed a rototiller, and planted a proper little patch of potatoes. We cut more trails and found hillsides of huckleberries just off the driveway. Sarah had been keeping bees for a couple of years now and had fought winter losses and collected a one-hundred-pound harvest of honey. Our herd of horses grew by fifty percent with the addition of Clyde, a handsome six-year-old Canadian. We had chickens and pigs, milking goats, and Tikka, Rose's German Shepherd.

From the time our girls were very young, we had normalized the raising and eventual butchering of domestic animals. We did not include them in the act of killing until they were older, but never hid any part of the process. The girls always helped skinning, gutting and cutting.

When Abigail was two, back on our first property, we made sure she understood this when we got the cute little lambs and wiener pigs. Later that summer, when we got our first barn kittens, Abby grabbed one and informed us, "We're gonna eat them."

Horrified, Rose replied, "No, Abigail, we're not going to eat them. We're just going to keep them."

Abby nodded in agreement and said, "I know – not now – but when they get big. Then we'll eat them."

We laughed, but Rose clearly felt like a bad mom, and she looked pained as she emphasized, "No, Abigail. Kittens are not for eating. We're not going to eat the kittens."

I touched her chin and turned her face to mine, adding, "We're not going to eat them Ab," just to drill it home.

Abigail gave us a big reassuring smile and said, "I know, we're not gonna eat them."

Rose and I shared a smile, feeling relieved.

And then Abby stretched her hands as far apart as she could and very seriously said, "But when they get BIG... then we're gonna eat them!"

Abigail and our first kitten

In 2016, we joined some friends on a trip to the Queen Charlotte Islands, now called Haida Gwaii, for some hunting and fishing. I had been there for the first time with Jay fifteen years earlier, braved the ocean in a canoe and shot little coastal blacktails with his .44 Mag. The misty coastal mountains full of game and seas rich with every manner of fish had left an indelible mark on my imagination.

Always pulling on my heart, we finally went again with Bev and Jay and their girls when Sarah, Abby, and Julia were four, two, and just a few months old. Sierra got sick on the long ferry ride from the mainland and we prayed it was just seasickness. We briefly explored the little town of Skidegate, bought our hunting tags and then drove to the remote west coast. We were still setting up our wall tent when Abigail first threw up on her PJs. We got our beds laid out on the ground and she threw up again, this time on her sleeping bag, highlighting our lack of running water and spare bedding.

While Rose worked some sort of mother-magic getting Abby cleaned up, we noticed a mouse sniffing around inside the tent. It didn't seem very concerned with us and, only when we locked eyes did it scurry back out.

The ferry ride had been seven hours long through the night and, by now, we'd been up for far too long. We were already cold, tired and now fighting sickness and cleaning up vomit. One little mouse was not going to bother us. But as Rose and I prepared our own bed, I noticed there were more.

We finally laid down and pulled up the covers. Rose and I lay face-to-face. I stared her directly in the eyes – dazzling blue – and wondered if she was mad at me or if she was super mad at me.

"You're amazing," I said.

She gave me a tiny, sweet smile and I continued.

"And gorgeous."

That's when I noticed the mouse. It emerged from the shadows behind her and quietly skittered to a stop just a few inches from the top of her head. Her beautiful, long blonde hair was fanned out on the ground above her and the mouse sat on its haunches right there on top of her hair. Suddenly feeling like this was a hostage situation, I slowly, quietly called.

"Rose...."

She looked at me like a drunk on the verge of passing out and just fluttered her eyelashes.

The mouse nibbled on something and stared at me. I looked at him and then back at Rose, who had a total look of indifference on her face. She did not like mice, so I thought she would've been displeased with one squatting on her hair. I couldn't see them, but statistically, there had to be turds by now. But she did not care.

"OK," I said, mostly to myself, and then reiterated my earlier sentiment. "You are amazing."

She closed her eyes, and so did I. If she didn't care about mice crawling on and around her all night long, neither did I.

The whole thing unravelled when the morning light revealed a mouse turd explosion, and Rose erupted.

"JEFF! Look!"

Her eyes darted about incredulously.

"Mice! They're everywhere!" she said, properly freaking out.

"I know," I said, confused by her sudden change of heart. "You didn't mind them last night."

"What!?" she stared daggers. "What happened last night?"

At this point, my survival instincts kicked in.

"Nothing, Rose," I lied. "We'll take care of them, don't worry about it."

We brushed turds off Sarah and Abby's sleeping bags, but it got worse when we checked Julia. Just two months old, we had brought a portable playpen, and when Rose went to fetch her, she cried,

"They ate her soother!"

I verified that; indeed, the end of the silicon soother had been chewed off and replaced with a sprinkling of turds on the baby blanket. A full-on war ensued, and one-hundred-and-four mice were killed in the next three days.

Sarah and Abby fished on the ocean for the first time in our little twelve-foot aluminum boat and caught their first rock cod. We shot some deer and ate surf and turf like kings. The whole adventure turned positive but never really recovered from its filthy origins.

Sarah with a Rockfish

We all packed into our little boat and bounced around in the open seas, catching rockfish, lingcod, and even a spring salmon. I managed to hook a huge halibut and battled it into the boat—a 55-pound sheet of plywood thrashing like a hydraulic bull. I had to jump on it and subdue it before it cracked our hull.

A gang of kids – our five and another five friends – explored the rocky coastline and combed the beaches. They even captained the boat by themselves. Haida Gwaii is a world-class outdoor adventure destination and is basically in our own backyard.

Julia with her first buck on the Queen Charlotte Islands

In 2017, after six years on the Tipper Creek homestead, we finally figured out the elk. Every year, we could hear them bugling right at the house. Every year, we tracked them and called them and saw them. Every year up till now, we were unable to get a shot at a legal six-point.

We hunted almost every morning through September and called in some nice five-pointers, but they needed the sixth point to be legal. One particular evening, we heard a bull bugling just a couple of hundred yards away down at the river. We could hardly sleep as he bugled through the night. At the crack of dawn, we wasted no time and headed straight to the water's edge. We quickly engaged with cow calls and he bugled back consistently but was hung up in the brush and wouldn't come out.

Jeff and a giant halibut in the 14' aluminum boat

Sarah was ready. She had my 7mm Rem Mag rested on a boulder and was pressed in amongst the rocks with the butt of the gun firmly on her shoulder. But the bull stubbornly refused to show himself, a sign he was probably a mature herd bull. I changed tactics. Instead of just cow calling, I bugled and started breaking limbs and smashing brush. He immediately responded and sounded angry.

"Get ready, Sarah," I instructed. With one more call, he pushed out onto the gravel bar just across the river.

"He's legal, Sarah, take him!" I urged.

She hesitated and the bull trotted a few yards and turned back to the willows.

"Shoot him now!" I insisted.

At the last possible moment, she pulled the trigger and made a direct hit. I was proud of Sarah but even prouder when the whole family of girls gutted, skinned, quartered, and loaded the whole elk onto the horses and packed it out.

Parenting is often a slurry of emotions – kind of like a stew, nutritious but not particularly exceptional. There are defining moments, however, when all the finest ingredients come together with just the right recipe. Sarah's perfect elk hunt served up like a filet mignon with sautéed prawns and tiramisu for dessert. I could never have guessed at the time, but I filmed and posted the video on our fledgling YouTube channel and that hunt would go on to amass millions of views.

We were definitely at the dawn of a golden age of family backcountry adventures.

Sarah, Abigail, and I had our first dangerous encounter when we jumped a mama grizzly bear and her cubs.

Later that season, with the whole family, we got snowed under in the alpine for thirty-six hours, snapping tent poles and freezing our boots solid.

We had a successful caribou hunt in the Rocky Mountain foothills and all the girls helped pack meat and antlers over miles of alpine and timber. We would go back to that same area a few years later with horses and shoot a giant bull moose. The girls had to struggle and fight to keep control of their horses on steep terrain without getting stomped on or knocked flying.

Still a hobby, I shared many of these adventures on our channel. Without much skill and in the total absence of strategy, the videos nevertheless slowly gained attention from outdoor and off-grid enthusiasts around the world. I didn't tell anyone about it except our close friends and family, yet our YouTube channel was about to pop off.

Jeff and Sarah and bull caribou in the Rocky Mountain foothills

The big girls, Sarah and Abby, shared a bunk bed built into the walls of their little six-by-eight foot bedroom. The little girls shared another custom bunk bed with a queen-size bed on the bottom and two-foot-by-five-foot upper bunks attached to the walls in an L-shape.

We knew this wouldn't last forever and, with Sarah turning 16, we suggested she should build herself a little house if she wanted more space. She always liked drawing little house plans and jumped on the idea. Our barn had never taken shape despite erecting the twenty-foot treated poles years prior. I proposed that she and Abby build a second-level house, leaving the ground floor open as a horse shelter and maybe, eventually, a milking parlour.

It wasn't her favourite idea but, with a foundation in place, they could get started right away. Over the course of two years, we helped Sarah and Abby fell trees, mill timbers, and frame up floors, walls, and a roof.

The walls were regular stud framing with Rockwool insulation, but for the 12/12-pitch roof we tried something experimental. We designed and built structurally insulated panels, four feet wide and twelve feet long. We took three-inch rigid insulation and applied plywood on the outside and 3/16-inch-thick pine veneers milled on our Wood-Mizer on the inside. Once we had the rafters and centre boarding in place, we threw the SIPs on top, screwed them down one by one, and the whole thing was buttoned up in a couple of hours.

I'm not a fan of tall houses or tall roofs, and I don't think I'd build one again. I'm not sure why I didn't learn my lesson completely on our first house build in Hudson's Hope with its thirty-foot-high ridge. It's a lot of hassle trying not to fall to your death. Safety came to mind after Abby dissected her pinky finger with an Olfa blade cutting foam for the SIPs.

Dave and Shoshanna lent us their old rock-climbing ropes and harnesses, and we anchored them to a tree on the opposite side of the house we were working on. They would hold if someone fell off the eaves, but not if someone fell off one of the gable ends. Sarah and Abby did most of the roof work while Dave and I supported from the gable ends.

It definitely locked in their sense of accomplishment and ownership. They milled board-and-batten siding for the outside and, for the inside, took that same siding, planed it smooth, and shaped it into shiplap with a dado blade on the table saw. They added a direct-vent propane wall heater to the cozy little living room downstairs and a couple of custom beds in the sleeping loft. They were only one hundred yards from the house, so they didn't bother with water, power or a kitchen.

Sarah and Abby lived there for a few years until Sarah moved out and Christina moved in. The cute little twelve-foot-by-twelve-foot tiny house on the hill stands as a testament to the girls' skill, hard work, and independence.

Living Like a Millionaire

Life was happening fast. I was transforming daily from businessman to homesteader and back again and the girls seemed to be growing up at an accelerating pace. It was a terrible wake-up call when the phone rang and a sobbing Sarah informed us that she had crashed and rolled her Jeep. She sounded like a wreck and we immediately hopped in the truck to find her and Abby. Just 20 minutes from home, we found them with puffy red faces, standing dejected and alone on the side of the road. The Jeep was lying on its roof and was smaller than the last time I'd seen it.

Miraculously, the girls were completely unharmed. I think fatigue was the major factor. Sarah had allowed the Jeep to drift off the road, and it rolled into the ditch and then flipped end-for-end. There weren't any skid marks; she never even hit the brakes. They were not far from the Schalins' house, and Ethan came with a tractor to flip the Jeep back on its wheels and pull it out of the ditch.

We weren't the parents of five cute little babies anymore. We had somehow landed squarely in the middle of an episode of The Wonder

Years. Realizing that our girls could actually grow up and move away, we suddenly had to ask ourselves some hard questions.

"Have we done everything we want to do with the girls before they're gone?"

Of course, the answer was no. We needed to do some more hunting and fishing and homesteading together. We needed to switch things up a bit.

Work at BBA had progressed as advertised. It was challenging and rewarding. I hired people. I fired people. I made a ton of proposals and even won some. My original team integrated very well and had now doubled in size. We had a lot of success in western Canada but never did gain traction in central Canada. There were failures that I felt bad about, but overall, I delivered on my original mandate.

It had been a fascinating two-and-a-half years and I truly enjoyed working with such a competent and committed team. With a good salary, but mostly due to the profitability of my shares as a partner, I was on track to make a ton of money. I was thirty-eight and if I stuck with it for another ten years, I could've stashed away millions. The money was tempting, but more tempting was just being part of a winning team. On the flip side, I didn't need millions of dollars and I was captain of a winning team at home.

The major difference was that the team at home was my first priority and had a rapidly approaching expiry date.

The homestead had come a long way from the untouched land we acquired eight years ago. It was getting very functional and downright comfortable, although there were still so many things we wanted to do. We needed more pasture. We needed more water for crops and animals. We needed bigger gardens and we desperately wanted to tie up all the loose ends.

What if, for the first time ever, I just took a full year off? Imagine what we could get done.

Our living expenses were really low with no mortgage and no utilities. We could certainly live off our savings. What if we spent our time and money on the property in ways that actually paid us back – like an investment? Could we invest $5,000 and a couple of weeks of our time in gardens and irrigation water and actually grow a thousand pounds of food every year? It was crazy that we could even consider it. We talked it through over and over, trying to find some good reason why I couldn't just quit and spend a year together as a family, hunting, fishing and homesteading. Our off-grid homestead had already paid big dividends, but it still had so much untapped potential. We had no idea how timely our homestead investments in 2019 would prove to be.

I stepped down and sold my shares back to BBA at the end of April 2019. I promised them, "It's not you, it's me." But no matter how you describe it, it was still a breakup. It was definitely surreal to sever such a good thing, but it was also a testament to how great our new life would be – or how great we imagined it would be. Of course, we had no idea, no previous experience and no good examples of what total unemployment looked like.

Those first few days waking up with no meetings scheduled and no emails to assemble felt like a spacewalk – no gravity and no ground beneath our feet. Even with the amount of autonomy I'd had over the last seven years, it did not hold a match to the formless expanse in front of me. With no boss, no responsibility to employees, no schedule, no predefined targets and no one to sanity-check my own goals, it became a challenge to even collect all my thoughts and ideas.

I can see why some people prefer the structure of employment, as the vast potential of full autonomy can be overwhelming. My mind was blown with possibilities. Slowly, I narrowed my focus onto the things directly in front of me. We had a homestead and we were gonna supercharge it.

The first major project was an expansive water system that would supply gardens, animal pens, and our future greenhouse. We installed a 1,000-gallon tank on the hill, which could gravity-feed every other location. With the Ditch Witch on the Toolcat, we dug shallow eighteen-inch trenches to both upper gardens, the goat pens, the greenhouse, the snacking gardens in front of the house and the duck pen. We only required water in the summer, so shallow lines would be fine and would be far cheaper and easier to install.

To fill the tank, we ran 1,000 feet of poly pipe to the beaver pond on our creek. Initially, we just used the Honda pump to fill the tank. We soon installed a ram pump that could automatically and continuously fill the tank all summer long. The ram pump only required about five to eight feet of drop from the beaver pond to the creek below and, using a water hammer effect, was able to pressurize the outlet and push half a gallon per minute up forty feet. It was almost miraculous to watch the steady dribble, day and night, keep our water storage full.

We also cleared half an acre of lowland below the garden and dug a giant pond, eighty feet long, fifty feet across, and ten to fifteen feet deep. If the ram pump failed, we could use a Honda pump to fill the tank from the pond. We could also use our new vast water supply to irrigate the gardens directly and flood our hockey rink in the winter. It served as a dugout for the horses, goats, and, eventually, a milking cow in a new four-acre forest pasture.

Our new riches in water had the expected effect and blew up our humble gardens into bursting red tomatoes, six-foot tangles of sweet peas, huge heads of broccoli and cauliflower, and our biggest harvest of crunchy apples.

We dug a proper little pond for the ducks and were quickly won over by their polite quacking, elegant preening and awkward waddling. The duck eggs were just like chicken eggs but a little richer in colour and flavour, just like their meat.

We built a ten-foot-by-twenty-foot greenhouse attached to the south side of our shop. Without any real topsoil, Rose successfully resorted again to planting directly in a bit of straw. That first year, we grew two-hundred pounds of tomatoes, jalapeño and sweet peppers, way too many pickling cukes and a surprising amount of basil. For a nothing-special location in Zone 2, we were super happy with our rookie results.

We built a second pigpen adjacent to our first and planted a big patch of potatoes in the original pigpen. Just a few years before, it had been a thick tangle of timber. We cut down the trees and let the pigs tear out all the brush and roots. In the fall, we collected all the woody debris the pigs had loosened and threw it in a huge heap to burn. Within a couple of years, everything except the biggest stumps was cleared out, and we had a quarter-acre available to garden.

We built small log-cabin-style raised garden beds right in front of the house. With poplars we had cleared from the yard, we simply notched and stacked them to build five-foot-by-twenty-foot beds about two feet high. Again, we didn't have any soil to fill them with, so we stuffed the bottom full of rotting logs and woody debris and then stuffed them to the top with straw. We dressed the top with just an inch or two of composted horse manure. Most things grew excellently, especially from starts, while a few others, like carrot seeds, struggled to survive through germination. Relatively small gardens and conveniently located right in front of the house, they were reserved for snacking foods – full of sweet peas, shelling peas, broccoli, cauliflower, and strawberries.

We continued to refine our shop and improve the guest house we had built in one corner.

We milled up a handful of large spruce and built a 500-square-foot shed to house our growing collection of tools, materials and equipment. Partially due to temperatures and conditions, but mostly because of the amount of snow accumulation, it's impractical to store or work on anything outside. If we wanted to use anything for the six months of winter, we had to build it a home with a roof.

Getting the stuff out of our shop gave us a lot more room for actual workspace. The front of the shop finally got dressed up as well with some river rock stonework and more cedar board-and-batten cladding. The stonework on the shop was my practice run for a short stone wall I would build behind the wood stove in the house.

The house got its share of other improvements as well. We put some finishing touches on the stucco walls where they met the ceiling joists and painted all the floors. We added some shelves to the pantry and built a little pony wall between our living room and kitchen that contained yet more shelving for all of Rose's preserves.

The biggest project for the house was a screened-in and mosquito-proof back porch. Mosquito season – or "mosquito hell," as we commonly referred to it – usually spans late May to early July but can be better or far worse, depending on the year. With the hot weather of spring arriving with hordes of ravenous little demons, it creates the perplexing dilemma of needing to go outside and also dreading going outside. The screened-in porch gave us a valuable outdoor living space and doubled as a summer kitchen with the addition of another wood cookstove.

The biggest build that year, for a couple of reasons, was the root cellars. For years, we had planned to build root cellars in the bank along our driveway. The scrap chunks of culvert intended for that purpose sat partially buried for five years. One culvert was nine feet in diameter and the other twelve feet in diameter, oriented vertically, open to the sky. The plan was to mill lumber and timbers and custom-fit them as forms to pour a suspended concrete slab for the roof. The front culvert would be only partially buried with a living roof and serve as the outer root cellar. The inner one would be completely buried with all the temperature and humidity characteristics of a true cellar.

It was a labor-intensive job, custom-fitting and supporting the formwork, tying all the rebar and then mixing and pouring the concrete. We capped the slab with torch-on roofing and then covered it with foam and dirt. The resulting product was an innocuous-looking little hobbit house nestled into the bank. They looked fantastic and worked even better. The other notable part of the root cellar project was that, when we posted the video, it blew up our tiny YouTube channel and kicked open a new door of opportunity.

Keziah and the root cellar

Sarah was graduating and she'd already secured a job as a dockhand at a prestigious fishing lodge on Haida Gwaii. She was also dating our good friends' eldest son, Ethan Schalin, who'd been trying to win her heart ever since they first met eight years earlier. Who knew how long she'd be back at home after her summer work adventure?

It occurred to me that for every year of my life, it was five child-years. That was a huge privilege and responsibility, and I was happy to trade one year of a rewarding career-dad for five kid-years of homestead-adventure-dad.

That year, we didn't just work on projects at the homestead – we took every opportunity to go fishing, hiking, prospecting, four-wheeling and hunting. We had just acquired a placer claim on the Fraser River and were finally sluicing a few grams of gold. We joined our friends on the coast to fish the ocean and did our best with our little fourteen-foot aluminum boat. We hunted deer back up in the Peace with Uncle Dave, and Julia and Christina both shot beauty whitetails. We explored new country in the Rocky Mountains on horseback, scouting and hunting with Aunt Joy and Uncle Shawn.

Sarah bought a new (old) Jeep Cherokee, and we jacked it up, cut the fenders, locked the rear diff, tabbed a front bumper and added a winch. We called it the Abominable Snowman.

Ethan and Sarah got engaged and had a proper, gorgeous summer wedding in the Schalin hay fields. We held a reception on the grassy lawn in our front yard that just nine years earlier was a brushy thicket. We BBQ'd wild salmon on the patio in front of the guest house. Rose served homemade, and very explosive, raspberry kombucha to all our guests.

Our inaccessible piece of cheap raw land had transformed into a bright and beautiful retreat in the wilderness. We also transformed into wiser, smarter, more skilled and free versions of ourselves.

Sarah and Ethan drove away in her lifted jeep.

Pulling a little tin fishing boat.

Our family adventure was not over, but already our kids were starting their own adventures. Just imagine five audacious girls, free of the matrix, each pursuing their own purpose!

It's hard for me to overstate how rewarding our unconventional journey has been. Unconventional, maybe even crazy (to the latte-sipping critics), but not impossible and not unreasonable.

You can do it!

Your journey won't be the same, and I'd advise you specifically to avoid the "45-gallon toilet juicer" part of our journey, but it could still be great We all have access to the important and simple things in life. We just have to throw out all the unaffordable complications that weigh us down.

Are you going Off Grid? Let us know!

You can connect with us at:

https://www.youtube.com/@Gridlessness

https://gridlessness.com/

https://x.com/Gridlessness

Epilogue

Grandad would've loved our little homestead. I wish he was able to see it. He could have taught us a lot about gardening and all the lessons he'd learned through a lifetime of making things better. I also wanted him to nonchalantly reach out and press a burning-hot teaspoon onto Sarah's unsuspecting arm. An endearing trick he learned from his grandfather – he would innocently stir the cream and sugar into his fresh cup of tea and then, with impeccable timing, withdraw it quietly from the cup and, without letting it cool for even a moment, press it firmly on the forearm of a young grandchild. Grandad's favourite target, I was familiar with the sting. It was sharp enough to make me jump as if hit by a bee. I knew the practice would skip a generation and ultimately only continue once I had my own grandchildren.

There would be many pivotal milestones in our off-grid homestead progression that I believed Grandad would have appreciated. Unfortunately for us, Grandad checked out a little early and, at the age of sixty-seven, abandoned his cancer-stricken body and went to be with the Lord.

Grandma was devastated and never recovered. A natural-born fighter, she survived into her mid-90s but, after Grandad died, wasn't really alive. She did, however, perk up and find strength whenever she visited our little homesteads. We loved those visits, filled with cups of tea, lots of books read to the kids, and endless puttering around the yard – organizing this and that and generally making things look better.

We continued homesteading and adventuring and sharing it all on YouTube. I was still not a YouTuber. I didn't know how it worked. I hadn't monetized the channel and I rarely watched other videos on the platform. While our videos had previously gotten a few thousand views, the root cellar video quickly went to 300,000 and, shortly after, a whole pile of older videos took off as well.

By early 2020, we had about 50,000 subscribers and were being contacted by every Chinese company who had ever manufactured anything, asking if we could promote their product. I didn't like that idea and also resisted monetizing by letting ads on. I continued to try to improve the quality of the videos, with the vision that my kids would watch them one day with their own kids. I would've loved to have been more skilled with the camera, editing and storytelling, but it was still just a hobby I did in my spare time.

It would be a few more years before I started to see my artistic outlet as something of a burgeoning profession. Just like I had been inspired by the footage and diary of Dick Proenneke recorded in the 1950s and '60s, there were a lot of people being inspired by our story. There were also some terrible, hateful trolls who threatened our lives and nearly had us abandon the whole thing. But the countless heartfelt messages won out. People shared their hopes and dreams with us and many relayed stories of how our videos had inspired them to take the plunge in their own lives.

One person reached out to tell us how much he and his mom enjoyed watching our videos. Apparently they reminded her of her own childhood. She grew sick and was in the hospital and had even lost her eyesight. For the last few days of her life, they watched (listened to) our videos. The sound of the girls' laughter brought her comfort and joy in her final days.

Wow! How's that for encouraging me to keep going and keep getting better?

Another time, a man reached out, sharing his appreciation for our family stories, especially for the girls and their horse-handling skills.

"blessings to all : luv your direction & ya sharing your doing the trip / journey .... Jeff & Rose - you guys are rockin it the way life should be > the gals ya got to be proud & they are cool too..."

He told us about a herd of perfect mountain horses he had laboured for most of his life to breed, and then made us an incredible offer:

"these ponies are out of 40 yrs of breeding.... come up from my upbringing & being a farrier for 20 + yrs & the horse industry was breeding for a lot of muscle on a fine & finer boned frame ; just fluff - show . I can appreciate but was looking to build a good strong well framed / good looking saddle - ranch horse's that can be rode / driven or packed > all in one > ... these horses were selected through the yrs ; crossed & its a labour of luv the see what ya get .... Bred for disposition - heart / mind .... If I may be so bold - we luv & look forward to you sharing your journey & respectfully think in my opinion - you guys are doing life & your gal's justice - we have to down grade ; the world needs more luv - I've seen where you guys go & understand the climate - these ponies are made to suit those conditions.... if ya want them to have & check them out they is yours ....."

Amazed by his offer, we emailed back and forth trying to find a way that worked for both of us. A herd of mountain horses would be awesome for sure, but fencing, feeding and hauling them all would be a major challenge. Just a month or two after our last exchange, we got the terrible news from his wife:

"Sadly, I must let you know that K---- passed away March 15th. I was well aware of his communication with you & was happy to see that he felt some hope for his beloved horses... I just wanted to update you on our happenings & would honour his wishes if you have any interest. Sending Love, Peace & many Blessings"

It was surreal to meet a stranger who knew us so well, to connect so quickly and then immediately mourn his death. A YouTube channel is perfectly impersonal, a parade of bits and bytes. Yet here we were, connecting to real people and kindred spirits from around the globe.

Sarah and Ethan eventually accepted the stud and two breeding mares, which have since thrown three phillies and a colt. With training from Sarah and Julia, we hope to take the young ones out on the trail soon. It gives me a sense of honour to help continue his legacy; to trot his beloved horses once again into the glorious mountain ranges.

We've heard from many young families who, after seeing our example, have chosen the path less travelled. For the sake of their faith, family and health, people are finding a way to do things differently.

One couple we came to know watched our very first videos in 2016 while still living in the UK. Wanting a better life and to start a family closer to nature, they immigrated to Canada, to our province of British Columbia. We eventually met them at a campout where they relayed their three-year ordeal of immigrating and moving, inspired the whole time by our family's story.

Even with a still-modest YouTube channel, we've heard from countless people trying to ditch the system and escape the rat race. I suppose this book is my response. We found a way, and so can you. Find that balance between material needs, purpose and productivity, and strong relationships. If you have a surplus of all three, then you've made it – you're a millionaire!

Regardless of your bank account.

A year slipped into two and then into three. With our homestead fully paid for, fully productive, and in pretty good shape, I wanted to work again but didn't need to work full-time. I contracted about five hundred hours a year – approximately quarter time.

At this point, we self-imposed a moratorium on large projects on the homestead – partially to keep our spending down, and partially so we'd have more time to adventure and just enjoy the peace, quiet and beauty on our private estate. Small projects were still "allowed." We built a wood-fired adobe pizza oven and forever transformed our fire pit cuisine.

We upgraded our fourteen-foot aluminum boat to a leaky old eighteen-foot aluminum boat that, Lord willing, would take us out onto the ocean. We rebuilt the transom, installed a pod, put in a custom fuel tank and rewired the whole thing. It still leaked, but it floated and opened up a whole new ocean of possibilities.

We were spending more time in the pay dirt down at our Fraser claim and finally amassed our first ounce of gold.

Julia was now a full-blown leather maker and was designing and building durable ammo belts, holsters, and other items for sale through our new website. I gave her my old Jeep Cherokee with a blown engine, and she bought a parts Jeep and swapped out the power plant.

Abigail graduated that year and, over the winter, we built a food truck. More specifically, she raised some money from unicorn investors, bought a twelve-foot enclosed trailer, and customized every inch of it into a fully-approved and functional poutine machine. By May 2022, she was pulling it around with her Jeep and serving the best poutine west of Montreal. She went on to work hard with her number-one employee, Christina, and pay it off in three years.

The years have blended together, punctuated by too many achievements, adventures, and picturesque moments to mention. It's now been fourteen years since we bought our raw land, thirteen since I started my own business, eleven since we became debt-free, and six since I worked full-time (for money).

We fished the ocean each summer for the last four years, explored secret beaches and cascading waterfalls, and caught enough salmon and halibut to fill our freezer and share with friends.

Christina holds up her Spring Salmon with help from Abigail, Julia and Keziah

Christina and I shot a tasty, little bull moose right by the house. The ability to hunt on foot from our front door and harvest wild, healthy meat is still deeply satisfying. The following year, we all went into the mountains and shot North America's largest animal – a beautiful bull bison. Last year, Julia shot the elk of a lifetime – a mature bull with a heavy, dark, long-tined rack.

We scabbed together a truck canopy for our trusty Ford, a boat rack and a couple of rooftop tents, stuffed it full of gear and explored the western United States and Mexico for six months. Encouraged by the people we met, the wild places we visited, and the productive spearfishing in the Sea of Cortez, we've made it an annual migration.

Rose and I have now been married twenty-six years, and none of this would've been possible without her. King Solomon summed it up: "Though one may be overpowered, two can defend themselves."

Our girls have all finished school and are now finding their own way. Thankfully, after spending so much time together as a family, we have many shared interests and hope the blessing of children will continue and increase. Indeed, Sarah has already delivered us three spirited little grandbabies – our true inheritance. Just a few more years, and I'll be patiently stirring my tea, just waiting for one of them to unwittingly sit down beside me...

Printed in Dunstable, United Kingdom